Reading Nine

About the author

The writer's personality should be rep-
resented in his books.

Born in Germany, to be precise, in Muelheim an der Ruhr, late on 26 October 1953 Joachim Endemann's desire to express his own opinion if something did not seem right — no matter if it was at school or in people's behaviour towards each other started in his early school days.

Then, with increasing awareness, during the time at secondary school (__advanced technical college certificate__), especially the discussions were contributed to articulating the own perspective: The only thing that counted was the better argument. This development continued during the vocational training (__as a *qualified window dresser, and as a physiotherapist*__) and during his studies (__primarily at *Open University: at first 4 terms at the Open University of Hagen* [__among other things *Social Sciences*__], then in the 1980ies 2 terms of *Art History with the focus on Iconography* at the University of Santiago de Compostela and, again years later, 4 terms of *Etudes germaniques* [__including *Histoire contemporaine allemande*__] at the Open University of the "Mirail" in Toulouse, respectively since 1994 with him often spending several months a year in his house in the French Pyrenees.

Among other things these studies were accompanied by translations which intensified his understanding of ranges of topics and their more or less well-done lighting up in the respective written outlines. For, as we all know, "History" is one thing, "Historiography" is another. That means history and historiography are two different pairs of shoes.

A major feature of his personality is that "climbing up the career ladder" has never been an issue to the author, but his focus has always been to get a better understanding of other people and himself as well: *Gnothi sauton* ("Know thyself") once was to be read — and for a good reason — at the Temple of Apollo in Delphi — in order to finally get to what *Anthropos* can mean, which, however, also involves knowledge about the conditions that are part of the prerequisites so that the potential inherent in any human being can actually develop. — And it is herein that the approach to understand his books published in the

edition !_*scheuklappenfrei*_! respectively !_*think out-of-the-box*_!
by EndemannPublisher lies.

So far the following volumes have been published within the edition !_scheuklappenfrei_! respectively edition !_think out-of-the-box_!

Volumes 1-4:

Es werde m e h r Licht! Mehr Demokratie wagen in der Lobbykratie? Untersuchung über die Konsequenzen der bürgerlichen Real-Demokratie, 1. Auflage, Juni 2016, 2., revidierte Auflage, April 2018.

Volume 5:

Zwischenrufe in satirisch-politischen Variationen oder Reale Betrachtungen dadaistisch-surrealer Phänomene in der Lobbykratie, 1. Auflage, Dezember 2016, 2., revidierte Auflage, April 2018.

Volume 6 (__6.1 + 6.2__):

„Ich stimme nicht zu!" — Gesellschaftspolitische Lesungen über den Neowilhelmoliberalismus und seine Konsequenzen, Januar 2018.

Editorial note:

As the mentioned-above books are all directly linked as regards the topic, they make up the volumes I to III of a "narrational scientific trilogy" titled The *Tri*_logical Dissection of the Lobbycratic Era.

Volume 7.1 (__special volume__):

The present special volume titled *Reading Nine* is a part of the upcoming volume III of The *Tri*_logical Dissection of the Lobbycratic Era — "'I do not agree!' Politico-economic Readings on the *neo*_Wilhelmo-liberalism and its Consequences". (__*The publication of the complete volume is scheduled in fall 2018.*__)

R e a d i n g   *N i n e*

of

The *tri*_logical Dissection
of the Lobbycratic Era

Volume III

"I do not agree!"
Politico-economic Readings
on
The *neo*_Wilhelmo-liberalism
and its Consequences

by Joachim H.E. Endemann

EndemannPublisher Edition !_*think out-of-the-box*_! _ volume 7.1

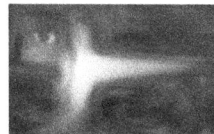

Reading Nine is a special volume as part of The *Tri*_logical Dissection of the Lobbycratic Era, volume III: "I do not agree!" Politico-economic Readings on the *neo*_Wilhelmo-liberalism and its Consequences.

This special volume is available as a PDF file with EndemannPublisher or is available as a book with Lulu Press, printed in the USA by Lulu Press Inc., Morrisville, NC

Typeset in 11 + 9 pt Corbel+

Layout by Joachim Endemann

© Joachim Endemann, June 2018

ISBN 978-3-9819719-2-7

This publication is listed with the German National Library (__DNB__); bibliographic details are available online at:
https://portal.dnb.de.

Worries about the respected reader by the author ...

In order to enable you to a better grasp, the dangerous trains of thoughts that waft through this volume of

The Tri_logical Dissection of the Lobbycratic Era

have a special layout structure that "saves" you, the reader, so that you, the reader, are left with the not unrealistic chance to get out of these trains of thoughts relatively unharmed — which actually does not say against the author, after all this structure harks back on his initiative.

To start off with

The concluding volume of the

> *Tri*_logical Dissection of the Lobbycratic Era

also features some peculiarities that will be presented in the following:

> With this volume — *for the first time ever* — trains of thoughts have been visualised.

Although first attempts to do so were made in the first German edition of volume I and II of this *trilogy* through three levels of text: the main text, the inserted text and the text in the footnotes, this was not sufficient as, on the one hand, the main text and the inserted text could not be sufficiently demarcated, and, on the other hand, the demarcation of the text in the footnotes proved to be too strong. This was particularly the case as the level of the text in the footnotes as of a certain extent start to become an alien element which ever and anon can make it difficult to see the context of the branching of a train of thought.

For this reason, this unique technique that was developed for the present book, subsequently was also applied to the revised editions of the first two German volumes of this trilogy: *Es werde mehr Licht! Mehr Demokratie wagen in der Lobbykratie? Untersuchung über die Konsequenzen der bürgerlichen Real-Demokratie,* as the volume I of the *Tri_logical* Dissection, and of *Zwischenrufe in satirisch-politischen Variationen oder Reale Betrachtungen dadaistisch-surrealer Phänomene in der Lobbykratie,* as the volume II of the *Tri_logical* Dissection. Both revised editions were published in April 2018.

Well, you have to admit that there is something fascinating about literally understanding the term 'train of thoughts' and about imagining a book as a continuum of trains of thoughts, that is to be expressed through this technique. Thus, it is not unusual that you will encounter complex compound sentences that are optically offset. This way there is a textual structure that makes the content of a complex structure of trains of thoughts accessible; especially through structuring layout elements. This is particularly the case as the use of these elements allows for intertwining a secondary train of thoughts that is especially offset and, thus, this facilitates seeing the context

without

confusing the course of the train of thoughts which would be the case if intertwining such a second train of thoughts remained without optically offsetting the main train of thought.

* * *

Another peculiarity of all the volumes published within the edition !_think out-of-the-box_! is the use of the abbreviations "B.O.Q.C." that is "before our questionable calendar", or "O.Q.C. that is of "our questionable calendar", as neither through the phrases "before Common Era" or "Common Era" nor through the other ones "before Christ" or "Anno Domini / A.D." a real fixed point of time is stated for the breakdown of the Roman Empire created gaps in the timeline thus calendrical uncertainty. And it is this uncertainty that makes terms such as "A.D." a matter of belief and is an attempt to suggest "clarity" and "security" where actually we can only have assumptions.[1]

---

* * *

Needless to say, that, just like in the German volumes of the Tri_logical Dissection of the Lobbycratic Era, the word (_"Wahlk_r_ampf" instead of "Wahlkampf"_) "election c_r_ampaign" (_instead of "election campaign"_) is not a misprint ...

---

* * *

---

[1] Mind you, it was only through the breakdown of the Roman Empire in the West, this was mainly as of 375 O.Q.C. when the Migration Period began and the local tribes were mingling with nomad groups from Central Asia that this calendrical uncertainty was disseminated. After all, we have no secured knowledge about this process, accompanied by calendrical confusion. — (_By the way the East Roman Empire did not transition into the Byzantine empire until the middle of the 7th century and it was probably also caused by warlike conflicts which sooner or later make _any_ great empire unruly._)

I would like to conclude these preliminaries with drawing your attention to three remarkable people.

Well, there is Professor Dr. Heiner Flassbeck whose publications enabled me to get a sound understanding what any other person per se seems to understand: the economy. However, I do not assume that Heiner Flassbeck will agree to all my conclusions. But is that what it is about?

As well, I would like to mention Professor Robert M Hayden (J.D., Ph.D.), who teaches at the "Pitt" of Pittsburgh in Pennsylvania, U.S.A., who generously allowed me to benefit from his expertise on the Balkans and the wars and conflicts in the 1990s, that destroyed the society of the former Yugoslavia.

Last but not least my wife Kirsten Grunau may not go unmentioned. After all, we share our lives, go together on our study trips and discuss and plan my works. Not only does she proof-read my scripts but, as she is an excellent Anglicist, she also translates the thoughts that I have clothed in German words, into proper English. — Hence, you, my dearest reader, now benefit from this first publication in English that you are holding in your hands!

This special volume only consists of the final reading and of
one of the annexes mentioned in the following

Table of Contents

of the volume III of

"I do not agree!"
Politico-economic Readings on
the *neo*_Wilhelmo-liberalism and its Consequences
of
The *Tri*_logical Dissection of the Lobbycratic Era

Preface

Any knowledge can only be fragmentary, thus forming your own opinion can by no means be complete. Those filters that cut out things and only allow for those things to happen that deliberately have been selected by others, hence have been processed to form opinions and this way suggesting to me that I actually had a "clear view" are not helpful but they are actually harmful.

Howbeit, forming your own opinion is accomplished best if you act as a human being as a whole, that is neither overly intellectual nor overly emotional, thus without blinders, well aware that the knowledge gathered will remain incomplete. Gathering knowledge without blinders means placing this knowledge in the corresponding context, cognisant that the process of gathering knowledge extends more or less deep into the unknown. But where the consciousness is in contact with the unknown, there are those "sensors" that once, at the stage of the first forming of knowledge, had formed, now, however, serving to associate what has just become conscious and what has already been known. —

Such an opinion formation must have preceded the development of an individual, needn't it? Therefore, the following question could be raised:

This is supposed to be feasible in a mass society? — Well, if it deserves the name democracy, it is.

QUOTATION

"The neoliberal project will not be politically compromised until a crucial demographic quantity of those people who understand that obviously nothing else can be expected from this project is reached."

END OF QUOTATION[1]

From my point of view, the situation is as follows: on the one side there is an elite that identifies itself with the lobbycratic EU and that, to a certain extent, has its effect way down into the European peoples through organisation as for example mentioned in *Reading Twenty* of the German version of The *tri*_logical Dissection of the Lobbycratic Era, volume III, part 2. But this way you cannot bind the mass of the people to the lobbycratic EU, in fact, this is only possible through force. Projects like the German *agenda 2010*[2], whose implementation leads to trim the mass of the people in line with the market, so that, in the end they can't help but function in the sense of this ideological guideline of the EU, thus in the sense of the EU ideology.[3]

---

[1] Source: Sebastian Müller, „Allianz des 'progressiven' Neoliberalismus", Makroskop.eu, 18 Mai 2017 at:
https://makroskop.eu/2017/05/die-allianz-des-progressiven-neoliberalismus/?success=1. This link was re-checked on 26 June 2018, the quotation was translated by the author.

[2] Please refer to the entry: "Agenda 2010" on Wikipedia: https://en.wikipedia.org/wiki/Agenda_2010. In the annex of the present special volume the main part of this "agenda" is mentioned on the pages 104-6.

[3] The ideologic objective of the EU is the „neoliberalism" which is dealt with in the German version of volume I of The *tri*_logical Dissection of the Lobbycratic Era. (_In the following mostly referred to as: The *tri*_logical Dissection [...], followed by "vol. I" or "vol. II" or "vol. III"._)

Resulting from that, a ruling class as cruel as can be is evolving, for if they were not that cruel, they would not ask for such a social trimming. On the other side there is the mass of the people, gullible and hoping per se, that is being led — by the corresponding political will-o'-the-wisps by means of rhetoric that is adapted to the very audience and the collective atmosphere via a political swing to the right or to the left but always back to the neoliberal direction. That at least is the approach that has proven well for the preservation of the power of the ruling class, always accompanied by the corresponding ballyhoo. By the way, this does not only work out so easily because the neoliberal doctrine has a very sophisticated theoretical construct of ideas at its disposal.

This is a construct of thoughts that I dare to call a dadaist-surreal one, which is, by the way not at all an offense to any upright or by now lying Dadaist or Surrealist — on condition the term Dada of the Dada movement is _never ever_ used in the context of the lobbycracy, but in this context the Dada of the lobbycratic era is to be called Dadaism — in contrast to the Dada of the „dada" movement, for „Dada" is _not_ theorisable. Neither is, in fact, the Dada of the lobbycratic era, but the term "Dadaism" is appropriate in this context, for it looks as if it were theorisable — and the same applies to the human being

> that according to neoliberalism is also theorisable, which, of course it is not.

It should be added that the neoliberal doctrine did exactly not come without presuppositions, it did not come out of the blue, kind of like a visitation and planned by a sinister power, but because of a behaviour in line with the market that psychologically has been grounded in the mass of the people in Europe long since.

> This can, amongst others, be seen in the fact that the marvellous term "time and leisure" has had a negative connotation for a _very_ long time, even though, building your _own_ opinion without time and leisure is in fact not possible.

After all, the industrialisation asked for forming such a corresponding collective mentality, which, may I mention, took place in a peculiar way in Germany and that led to, what needs to be called Wilhelminism. Interestingly, nowadays this mentality is coming back,

> (__well after a certain lead time that started, at a rough estimate, with the end of the Cold War __),

and it is coming back in a form that may well be called neo_Wilhelminism and that is currently amalgamating with the Neoliberalism to the phenomenon of neo_Wilhelmo-liberalism which is the base of what is currently shaping in the EU and that goes hand in hand with a kind of *neo*_imperialism. And, objectively seen, it is this monstrosity that the EU elite and its organisations, sometimes called NGOs, with their effects way down in the societies of the EU identify with.

> Mind you, it absolutely does not matter, if the members of these organisations realise that or if they indignantly want to deny it, for objectively seen you cannot deny it – on condition that you look at what has been happening ever since the end of the Cold War in an out-of-the-box manner without being ideologised.[4]

This process of neo_imperialism based on a neo_Wilhelmo-liberal regimen, will be _relatively_ completed by 2020, when an EU army will have been constituted, no need to say that it will under the rule of the German hegemon.

> This is a hegemon who would completely lack his own substance, as ever, if it were not for the mis-constructed EMU.[5]

However, first of all this army will be busy preserving the neoliberal peace and quiet — well supported by the smooth propaganda of the writing staff of the media concerns. Just to make sure that everybody does understand, when, (__for example__) Greece needs to be occupied or that in other regions of the EU social movements need to be suppressed, movements that not only confront themselves with intentions by the EU that affect them _directly_ and that finally even hold a referendum, but that eventually vote against lobbycracy, which then officially would be called anti-democratic, as it

---

[4] See in the German version of The tri_logical Dissection [...], volume III, part 2, Reading 20.

[5] You will find details on the "lack of own substance" further on in this copy.

happened in Wallonia in October in year 5 of the lobbycratic era when so-called great German democrats — called that vote a vote against democracy — that means that German politicians and so-called *alpha* Journalist vilified a decision-making process that was in line with democracy and anti-lobbycratic as "anti-democratic".

> Once you have read the volume III of the German version of The *tri_*logical Dissection of the Lobbycratic Era you will definitely realise why the year 2016 of O.Q.C. has to be year 5 of the lobbycratic era.

Well, in fact, the mass of the EU inmates is best to be held at bay — that means apart from the common mawkishness such as the national anthem and so on — by spotting an opponent that is not within the EU and, at best, can be _strikingly_ personalised: At present there would be Putin, Assad, Trump, Erdogan — ever changing accordingly to the own hypocritical requirements. Thus, we are talking about opponents in whose sphere of influence it _may_ seem appropriate to initiate an "export of democracy" to — either covertly or, in case the opponent is still a too major one (__thus for the time being__), and/or accompanied by proxy wars, which, of course will not be called so, but (__due to the own per-se-being-good__) these wars are called "human rights wars".

> The Orwellian creation of the word "human rights wars" is especially important for the salvation of the pacifists' souls, so that they can put their minds at rest and, nevertheless, participate — when it is about waging wars to expand the spheres of the ruling class of their so-called "own" national state.

> > You see that the neo-imperialism primarily differs in a semantic way from the paleo-imperialism.

It is self-speaking that even those ones will agree to all this (__*if they are not the first ones to do so*__) in parliament, who have never been in the army, let alone their children — and that means above all members of the so-called Green Party.[6]

At any rate, the lobbycratic era does not allow for any _constructive__ social ways to develop, due to the intention of its ideology that urges for homogenisation, even though there is permanent talking about "individuality" and "freedom".

> (__Of course, these are terms that everybody has to like, but, because of the social trimming that has taken place for long these terms themselves are also thought in a way that is in line with the market, thus they are actually thought as a reflex which in the end means they are not thought at all, merely used as a "knee-jerk".__)

Because of all this, in fact, a revolutionary situation might actually arise. And what would that mean? Well, it would probably mean what the later door-opener of a certain Mr Hitler stated as early as in 1912:

> "Either we will have a revolution in three years or we will be at war."

Mind you that with these words Emil Kirdorf only expressed the disposition of the European power elites — for on the territories of _their_ European playgrounds (_commonly known as national states_) strikes were the daily fare.

---

[6] See e.g. in the German version of The *Tri*_logical dissection [...]*, volume I, Chapter 14.

> Is it worth mentioning that the power elites went for war?

However, it is less likely that such a situation, possibly developing in a revolutionary way might really result in a revolution, made by the mass of the people — for there is no revolution without the _subjective_ factor.

---
* * *

Thus the intention to write these *Politico-economic readings*[7] is twofold: Firstly to serve the enlightenment of the sympathetic reader who is interested in his own interests that basically are different to those of the power elite of each national state and their satellites in *spin_doctorial* sciences, politics and journalism. That, secondly, implies these *Politico-economic readings* are also addressed to all those intellectuals, journalists, cultural workers as well as to all those who defend the neo-liberal politics wherever, that is to say, with regard to Germany: thus, these *Politico-economic readings* are a polemic echo to all those who defend a policy animated by the mother of the Merkel-ism and by the inventor of the Schäuble-ism

> (_i.e. the names of the persons currently acting may be changed, whereas the neoliberal direction is maintained_)

as the basis for the occurring homogenisation of all the societies in an "iron cage of austerity" called EU

---

[7] ... based on the German version of volume III of The *tri_logical* Dissection [...].

> (__instead of aligning the market to the societal claims as it were characteristic for a democracy__)

together with the yonder politicians of so-called left or so-called progressive parties, even reaching down to so-called bourgeois action groups of market conditioned younger people, who all show by their political demeanour and action that, in fact, they do not want anything else than supporting the lobbycratic policy!

> It follows from the above that all those who work or behave in the manner mentioned are responsible for the present development and its consequences.

Hence, if voters have the impression there is merely the choice between the present politically wrong direction and another politically wrong direction, resulting from the present one, they vote for the "new" wrong direction with the aim to abolish the present lobbycratic direction — and then they are betrayed again, fooled by another political ballyhoo.

Thus, if you feel the need for a real politico-economic change you cannot help but realise that first and foremost enlightenment is required — before

> (__with some prospect of success__)

any decision followed by any action for change is to be envisaged.

Joachim Endemann
Il Piano
May 2018

R e a d i n g   N i n e

The destruction of the European unification process by neoliberalism and *neo_*Wilhelminism

A sociopolitical European unification process in a representative democratic way can be successful *_only_* on condition that it does not happen under false pretenses. Transparency with *_all_* political decisions determining directions is required to ensure its successful implementation.

> Whereby politics, science, the mass media and the creative artists as catalysts all are assigned with crucial roles.

What does "European unification process" mean for a democratic polity?

❖ In this process, the main part of the human sciences is drawing up the required roadmap.

> That is including the definition of reference points where the process must come to a halt to reflect on the road

taken so far and to determine its con-
tinuation and the consequences resulting with a potential change of direction
to the next reference point.

❖ The mass media are hereby assigned with the task

(_serving as eyes and ears of the mass of people_)

to ensure the transparency and publicly announc-
ing this process —

supported by the creative artists,

❖ whereas politics must implement this process.

... decision-making in a representative-democratic way
has not taken place any more for long...

Reality lacks all the points mentioned above — on the contrary: decision-making in a representative-democratic way
has not taken place any more for long — due to the mere facts

❖ that all members of parliament are more or less
surrounded by lobbyists,

❖ that the human sciences are pegged into institu-
tions and think tanks ideologically oriented,

❖ that the journalists fancy themselves in their attempt to implant a

"neoliberal feeling of belonging together"

into the minds of the mass of the people of a society that is trimmed in a neoliberal way by news that they themselves have censoriously cleared and that are now called

"coverage oriented to the public welfare".

By the way, mawkishness is not an expression of public spirit, but it is an expression of emotional suffering due to the feigned attitude to life:

"we are good".[1]

Intrauterine censor-chip implantation

This kind of coverage is equivalent to the self-censoring phenomenon "scissors in the head" which means self-mined censorship that conveniently can take place right there.

However, it seems that this laborious and actually antiquated method is only going to be applied in the transition period for once the technological know-how can be applied successfully to implant intrauterinally a censor-chip as part of the future common ovum-tuning, applicable to all foetuses — well, at least to those who are later on meant to be politicians, spin-doctorial scientists, journalists and creative artists —, so that, among other things, the trimming of the society that has to be in line with the

---

[1] See the German version of The tri-logical Dissection [...], volume III, pp 358-9, beginning with: "Das Ausmaß der Projektion" (__meaning: "The extent of the projection"__).

> market corresponds to a *convincing* homogenous coverage, that is currently not perfect yet — at least there are several contemporaries who would call it insufficient ...

Initially, this "coverage oriented to the public welfare" was referring to Germany, thus was applied in the sense of the German unification in the 1990s but was then transferred to the European Union (__EU__) level and then was and has been practiced in the sense of the neoliberal trimming of this union, thus the trimming process of the whole European Union in line with the market.

❖ Those creative artists that depend on public orders respectively those who see themselves as creative artists actively involved in the "cultural sector" see their mission the same way as the latter ones.

❖ German politicians see themselves per se as door opener for the export economy and would therefore "never do anything against the economy" — just to quote the former chancellor Gerhard Schröder, who, with these words, exemplarily expressed that the proper function of an economy as a whole is foreign to those politicians setting the directions, not to mention their obvious non-understanding of the functioning of a monetary union.

Thus, objectively seen ...

> Besides, "objectively seen" is, in this context, to be understood as follows:

> observing political processes "out-of-the-box" with the proper distance to those causing political things happening — this way avoiding the danger of being personally taken in.

... Thus, objectively seen, those politicians could never deliberately do anything for the economy this way as long as they eagerly give in to the smoothest microeconomic interests instead of confining themselves to setting a well-designed macroeconomic framework — an elementary part of which the "Golden Rule of Wages"[2] would have to be.

> So, how could those politicians possibly do anything for the public well-being, domestically or internationally?

* * *

---

[2] See Reading One of The *Tri*_logical Dissection of the Lobbycratic Era, Volume III: "'I do not agree!' Politico-economic Readings about The *Neo*_Wilhelmo-liberalism and its Consequences" which will conclude with the current Reading Nine. Unfortunately, this complete volume is not available in English, yet but is currently being translated into proper English and will be published soon.

Excursion:

Quintessential examples of counterproductive consequences of the German machtpolitik

In the context of the responses to the Brexit and the election of Mr Trump on the part of EU politics which is determined by the German hegemon somebody uttered that it was now about the common interests of Europe.

> But what exactly are these common interests?

Well, that is not the question _for_ the actual question has to be:

> What are the interests of the power elites of Europe — respectively what are the interests of the German power elite?

After the end of the Cold War the common European interest could _only_ have been to pursue completely different politics, as a convincing expression of having drawn the consequences from the causes and the results of the two parts of the great war of the 20th century — politics which would have been possible if especially the German power elite and their satellites in spin-doctorial science, politics and the writing staff of the media concerns had been willing to do so. However, the first political decision from the German side affecting Europe was

> using that grain of sovereignty

to destabilise the situation on the Balkans which had traditionally been a difficult one by an irresponsible and rushing ahead policy.

QUOTATION

> [...] The Federal Republic of Germany has dictated the speed in offi-
> cially recognising Croatia without thinking about how to create a
> free, new sovereign state that has to deal with its minorities. [...]

END OF QUOTATION[3]

Solo action including blackmail

> On 17 December 1991, at a confer-
> ence of the then EC foreign minis-
> ters the German government

> (__that had priorly been urged to recognise Croatia and
> Slovenia by the opposition parties SPD [__the so-called
> Social Democratic Party__] and the Greens as well as
> the major German opinion-making "quality media"__)

> linked their "willingness to acceler-
> ate the European integration, most
> of all the creation of a common cur-
> rency" with "enforcing their way
> regarding the topic of Yugoslavia".

Usually this is called blackmailing.

> The result was then, on 15 January,
> 1992 Croatia and Slovenia

---

[3] Alfred Grosser in: *Die Woche*, issue of 13 July 1993; own translation.

> were officially recognised by the EC member-states, with the German government having them already recognised on 23 December 1991.[4]

End of the annotation: Solo action including blackmail

After all, the power relations at that time were suchlike that confederal structures could have been created with the main members of the EU[5], the USA, Russia and the UN, if especially Germany had taken the stand for exclusively that solution, expressing:

> We have learned our lessons of the two parts of the great war of the 20th century.

However, objectively seen, not relying on lip services paid but in terms of its results, today we have to draw the conclusion that none of those ones responsible in German politics, social sciences and media have understood anything about these lessons from history —

> let alone having learned _any lessons_ at all.

The power structures at that time would have allowed for creating a political entity on the Balkans involving the main states of the EU, the USA, Russia and the UN through a long-term approach based on the ambitious peace plan developed and

---

[4] For quotations regarding this annotation please refer to the quotation on p40.

[5] Respectively the European Community (__EC__).

submitted in 1992 by Robert M. Hayden, the anthropologist who teaches and researches political anthropology, law and international affairs at "the Pitt"[6].

Reflections on the consequences of
Yugoslavia's destruction

Hayden's plan can be summarised in the following seven points:

1. Abjudicating of the independent Yugoslavian republics and revocation of their representatives' participation in international bodies.

2. All military and paramilitary entities are to be assigned to direct control by the UN.

3. An overall-Yugoslavian interim government is to be appointed by the UN

> (__comparable to the post-war interim government in Germany after the end of the second part of the so-called Great War__),

exclusive of the members of the nationalist, chauvinist governments of the former autonomous republics

---

[6] The University of Pittsburgh, commonly called "the Pitt", is one of five universities of this second largest town of the US state of Pennsylvania, counting 300,000 inhabitants, following Philadelphia.

of Yugoslavia that were elected in 1990. This interim government reports exclusively to the UN.

4. Radio- and TV broadcasting companies are to be assigned to UN control, as the channels are pure propaganda and even outperform those ones broadcast in the Nazi era.

5. A committee of foreign experts on constitutional law is to draw up the constitution of a new Yugoslavian confederation. This constitution is to create a balance between developed autonomy and power of the central government, and it is this central government that has to be in charge of defense, trade, communication and the protection of minorities.

6.

Once this political entity has been put into operation and the monitoring of all the autonomous structures as well as those ones concerning the central government structures has been implemented and ensured, after one year a referendum, whether the respective populations wish an independent state or wish to remain in the new federation is to be held on the district and county levels. Should the people decide for an independent state an international commission will fix the borders of any newly created state. Should establishing new borders mean that a district remains isolated, there is to be a second referendum on this district. Should, however, the population in this district confirm their decision a second time, they are to be given full autonomy within the borders of that district.

7. For all states arising from this referendum member-
    ship in a military alliance is mandatory.[7]

---

* * *

Certainly, the conflicts between the different ethnic
groups had been there before, but it is not true, that there
could actually have been a guarantee for the safety of the
minorities, as the members of the governments of the
newly constituted republics were nationalist respectively
chauvinistic: the authorities behaved in a discriminating
and provocative way towards the minorities just as the TV
and radio stations agitated in a racist way for all this was
no coincidence but is owed to those responsible in the
governments of these republics — and thus also to the
Croatian and Slovenian government supported especially
by the German side. This entails a direct German co-re-
sponsibility for the escalation resulting in war and destruction by exactly this rushing ahead in recognising Slovenia
and Croatia which was, to make matters worse, combined
with the extorted demand from the EC member states to
accept this German policy for Yugoslavia, for otherwise
there would be no progress with the European integration. As a consequence, creating a common currency
would not have been on the agenda any more.

---

[7] See: „Ein Friedensplan für Jugoslawien" by Robert M. Hayden, in: *Die
Zeit*, issue of 7 August 1992. (__This plan is available in the online archive of
Zeit online on the following link which was re-checked on 20 March 2018:
http://www.zeit.de/1992/33/ein-friedensplan-fuer-jugoslawien/komplet-
tansicht?print.__)

QUOTATION

[...] The Federal Republic of Germany had made their willingness to accelerate the European integration, particularly the creation of a common currency, dependent on enforcing their way on the topic of Yugoslavia. Recognising Slovenia and Croatia was followed by the declaration of independence of Bosnia-Herzegovina which was against the will of the Serbian ethnical group. This way the German politics for Yugoslavia crucially contributed to the escalation of war of the conflict. As an arsonist whose performance was truly worthy of world-champions the German diplomacy was completely overchallenged when the pacification of the conflict was due. Washington took over and the German "partners in leadership" were knocked a peg back to the second row. [...]

END OF QUOTATION[8]

---

[8] Source: Werner Pirker, "'Deutsche Brandstifter', — Vor zwanzig Jahren hat die Bundesrepublik die Anerkennung Kroatiens und Sloweniens durch die EG erzwungen", in: "jungeWelt", issue of 14 January 2012; by courtesy of the editors of *jungeWelt / Verlag 8. Mai* GmbH. (__Own translation.__)

> Western European thinking is still saturated with
> reflection on "nation" and "state"
> particularly based on ideas by Hegel

While the minds of the western Europeans, especially the Germans', are still saturated with the belief,

> the roots of which go back especially to Hegel's reflections on "nation" and "state",

that "democracy" and "state" and "nation" formed an entity, this way meaning that "democracy" be present to each "nation" per se —

> and a "state" be the expression of it.

Consequently, in a state there would not be a power elite with their own interests that differs from the interests of the mass of the people, that, on the contrary this elite would have an eye on the "common good"— defined by their satellites.

> Well, this is a kind of believing that is proven wrong by the facts.

And, following this reasoning where a state is imagined as a being, "minorities" would be expressed through fewer rights and by minor cultural relevance for they would be the "others" or the "rabble", if not a potential threat for "peace" and "democracy" within and between the "national states".[9]

Thus, in contradiction to this merely intellectual concept of the state

(__thus in _living_ contrast to what is thought and ideologically rooted, particularly by those west European intellectuals__),

Yugoslavia was the _living_ proof,

and this way in contrast to those thoughts mostly shaped by automatic think patterns such as "Weltgeist" (__"world spirit"__) or "spirit of God",

that the meanings of "democracy" and "state" are only assumable if they are used for the _whole_ population of a country.

That means the members of a country might be different in their own perspective of philosophically or religiously ideological worldview and by regional customs,

as, even though not compelling but well useful means for living together in peace, and for personal, societal and economic prosperity,

but

all

are part of a body politic run by democratic representatives

---

[9] For the issue of the national state see the German version of The tri_logical Dissection [...], volume I, chapter 15: „Menschenrechte, Völkerrecht und das Konstrukt des Nationalstaates" corresponding to "human rights, international right and the construct of the national state". Also refer to pp 593-4 of volume III of this German edition.

based on democratic legislative texts, organised and executed by what is

then

called a "state".

Certainly, this is an idealisation of what was to be realised in Yugoslavia, however, the tendency to do that, is significant —

considering all the miserable conditions of the result of the great war of the 20th century as well as the results of the Cold War.

Against this backdrop the political nit-pickers of the petty bourgeoisie in Yugoslavia were supported

in their ideas that were as queer as they were blinkered

by those politicians of the national states,

with the German politicians in the front,

and against _this_ backdrop I have no choice but to call them the biggest hypocrites in the world,

for these strange and blinkered ideas

meant dividing

what could only be divided by tearing it into pieces —

namely the traditional cultural unity in diversity of the peoples of the former Yugoslavia,

which was exactly not a "failed experiment" of inter-
connecting peoples that are hostile to each other —

as it was claimed.

Consequently, this support was a vital mistake

because these ideas

of "people", "volksgeist" and "state"

filled with prejudices

that eventually are an expression of the attempt to give a new, but substantially old con-
ceptual base taken from the feudal ideology

(__and merely adapted to the needs of the bourgeois power elite__)

to a social structure that had been homogenised in a historic region in a long process[10]

are not approved by the facts,

whereas experience proves
that in those areas where the
Yugoslavian peoples lived

---

[10] Regarding a "social structure that had been homogenised in a historic region in a long process", please refer to the following passage on p402
of volume I of the German version of The tri_logical Dissection [...], beginning with: „Die Organisierung eines historisch alten Volkes", meaning: "the
organisation of an historic ancient people".

> intermingled after 1945, they actually
> inter-married in large numbers.[11]

The same applies for the so-called Serbo-Croatian language because this language and its diverse dialects were commonly accepted by native speakers, so that the insistence on linguistically purging this language,

> from its Serbian elements (__*from a Croatian view*__)
> and from its Croatian elements (__*from a Serbian view*__)

is just another expression of the political nit-pickers' thinking about the "volksgeist" regularly entailing severe political consequences.

> Furthermore, doesn't the popularity and the support of the majority of the Yugoslavian peoples for the last prime minister of the Socialist Federal Republic of Yugoslavia Ante Marcović (__1924-2011__) show that the majority of the people there would have greatly appreciated a federal continuation of Yugoslavia?[12]

Consequently, after the end of the Cold War the wisdom of German politicians in charge was absent.

---

[11] See Hayden, "Imagined Communities and real Victims: Self-Determination and ethnic Cleansing in Yugoslavia", in: American Ethnologist 23 (4): pp783-801, American Anthropological Association, 1996, pp788-90.

[12] See idem: *Blueprints for a House Divided — The Constitutional logic of the Yugoslavia Conflicts*, The University of Michigan Press, 2000, pp2-3.

In fact, as absent as it has always been —
not considering occasional exceptions.

Thus, the German sovereignty means international trouble
for, up to now, it has not contributed to any constructive solutions when it comes to matters involving other states.

Therefore, as we have learned from experience dealing ad-
equately with sovereignty is hardly manageable to the majority of those Germans that are actively involved in politics, culture and — mostly — in journalism because of political immaturity and immoderateness.[13]

That means

surging ahead in recognising Slovenia and Croatia must be
understood as an expression of old Wilhelminist reflexes
shown by German politicians in charge, regardless
whether in opposition or in governmental responsibility,
as well as by those involved in culture and journalism, as
they did not realise the fact that especially the peoples in
the former Yugoslavia were interconnected and intermin-
gled so closely, that, obviously, not only could they live together but because of the "political reality" they actually
had to live together.[14]

And who knows what else will come to light

(__at least when you look at the result__)

---

[13] See Werner Pirker, "Deutsche Brandstifter", for further details see
footnote 8 on p40.

[14] Cf. Hayden, loc. cit., p142.

when it comes to the true answer to the question why, in the middle of 2015, Mrs Merkel,

| without agreement with |

> the other governments of the EU member states,

| kind of out of the blue, |

> decreed a cynical what she calls Flüchtlings-Politik which means translated "refugee policy" but in fact it means politics at the expense of the refugees, thus politics that has to overstrain everybody?

| Indeed, this cannot have been pure philanthropy. |

> For shortly before that, this person dared to throw into a refugee child that "Germany cannot take you all" on air, but it was not at all about stating this banality — it was _exclusively_ about consolation.

| Well, finding an answer to this question might have to be seen against the backdrop of those old German imperialist reflexes that in the beginning of the 1990ies caused misery and brought chaos to the people in Yugoslavia? ... |

Is Franjo Tudjman
not comparable to Slobodan Milosevic?

The nationalist, racist and anti-Semite Franjo Tudjman (__1922-1999__) was the first President of Croatia from 1990 to 1999.

> That means initially he was the president of the "Socialist Republic of Croatia" (__30 May 1990__), then, on 2 September 1992, after the declaration of independence, he became the president of the Republic of Croatia.

It is during this period that 400,000 Serbians were forced to emigrate and as Tudjman was also the Commander-in-Chief of the Croatian Armed forces, it is unlikely to assume that this emigration occurred without his being informed. Sources say that the Tudjman government degraded the 600,000 Serbians living in Croatia from second constitutive people to minority.

> But, either way, before that Serbians accounted for about 14 percent of the Croatian population, today this percentage is with 4 % close to nothing.

At the same time, the necessity to secure resolutions regarding nationalities was abolished with a two-thirds-majority by the deputies of the parliament of the Republic of Croatia.

The consequences arising from this

> did not only have an impact on the job lives of the Serbian population, but they could also be seen in the both provocative and brutal behaviour of the Croatian state bodies.

Furthermore, this also brought back memories of the Fascist "Ustaša"-state[15]

| who was a faithful vassal state to the Nazi regime from 1941 to 1945 |

and who passed racial laws following the Nazi pattern particularly directed against Serbs as well as Jews, Romanies and anti-fascists, who were locked in concentration camps and from where they occasionally were transported to labour camps in Germany or physically annihilated.

Mind you,

there were more than 20 concentration camps in this state the biggest of which, Jasenovac, which was also called, unfortunately quite rightful, "Auschwitz of the Balkans", because the main task of this concentration camp was the physical annihilation of the Serbs in Croatia.[16]

That means the characteristics, that especially the German and the US-American side attributed to the Yugoslav, respect-

---

[15] "Ustaša" can be translated with "insurgents".

[16] Source: "The Jasenovac Extermination Camp". The following link was re-checked on 1 August 2018: http://www.holocaustresearchproject.org/othercamps/jasenovac.html. Also see: „Kinder-KZs der kroatischen Ustaša-Terroristen im Zweiten Weltkrieg" (__meaning: "Concentration camps for children by the Croatian Ustaša terrorists in the Second World War"__), whose link was also re-checked on 1 August 2018:
http://www.zukunft-braucht-erinnerung.de/kinder-kzs-der-kroatischen-ustasa-terroristen-im-zweiten-weltkrieg/.

tively the Serbian President Slobodan Milosevic (__1941-2006__), who was, among others described as the "monster" or the "Hitler of the Balkans", were not applied to Franjo Tudjman because he functioned in the sense of the "West"?[17]

Playing your own responsibility down and obscuring it

In the time from 1989 to 1991,

> the time during which Tudjman founded the "Croatian Democratic Union" (__HDZ__), was elected President in the following year and the referendum regarding Croatia's independence took place,

Tudjman visited the Federal Republic of Germany on several occasions and talked to representatives of the then Chancellor Kohl government. And it was this government that, apart from Austria, recognised Croatia's independence first

without

any agreement

> just to make sure that "agreement" is not mixed up with "blackmail"

with their partners (__?__)

---

[17]See especially Misha Glenny, *The Balkans: Nationalism, War and the Great Powers*,1804–2011, Penguin Books, New York, 2012, which is the revised and extended version of the book: *The Balkans: Nationalism, War and the Great Powers*,1804–1999, which was published in 1999.

and

without

any guarantee for the rights of the minorities in Croatia.[18]

It be repeated:

In Germany the government was not alone with this policy, for not only were they supported by the opposition parties SPD and the Greens and by the established German media, but there were especially voices from within these political and media ranks, who downright asked for such a policy that is bound to lead to chaos.

Of course, this was for purely humanitarian reasons, just like the war of aggression against the Serbs later on was, too, wasn't it?

And this way all of them were

(__seen from the result__)

united in Wilhelminist reflexes.

Against this background it may be "understandable" that this irresponsible policy is either not mentioned at all, played down or obscured by the German side.

---

[18] See the quotations on pp35 and 40.

A rhetorical question

As in 1999 the "West" was able to fight a war of aggression against Serbia and bombed Belgrade, it must be allowed to ask the rhetorical question why in the beginning of the 1990ies NATO and the UN did not take the chance to intervene — in order to spatially separate the different nationalist parties? Against the backdrop of the state of facts that the population on the Balkan Peninsula had been multi-ethnical for long and that the nationalities question had been one of the "most suitable" reasons for the outbreak of the first part of the so-called Great War and that this question remained unsolved and only frozen _because of_ the Cold War, it should have been the primary task by the international community

> (_if there was true interest in the concerns of the mass of the population_)

to ensure that

> _exclusively_

the peoples on the Balkans through their representatives

> (_and that means _not_ through nationalist nit-pickers_)

should have got the chance to choose and to decide, if they, with regard to the multi-ethnicity that has been there for centuries, preferred to live in a great multi-ethnical Balkan confederation _or_ in several smaller federal units

> (_particularly if they preferred to continue in a Yugoslavian republic_)

or if they preferred to live in states _relatively_ ethnically separated with guaranteed rights for minorities

> _or_

if they in future wished to live in states ethnically separated — thus in states where political nit-pickers call the tune.

> The latter case, of course, would have meant collective forced relocations of ethnic minorities because such a "sharp separatist referendum" would have been a hint that a peaceful living together in a common state would not seem probable in the future.

Actually, due to the severeness of the nationalities question and its, as potential as predictable, consequences

> (_if the floor is yielded to the political nit-pickers or they are even encouraged_)

such a procedure would _not only_ have been absolutely justified but it would have been a necessity — especially for an EU that was on the brink of its constitution at that time

> so that we could today rightfully say:

> "The EU has prevented wars in Europe".

Robert M. Hayden's plan mentioned at the outset of these

"Reflections on the consequences of the destruction of Yugo-slavia" would have been most appropriate as a roadmap whether "merely" for a constructive solution oriented at the interests of the mass of the people that live absolutely inter-mingled in many parts of the former Yugoslavia or even for a great Balkan confederation under the roof of the EU, with a simultaneous establishment of a security partnership includ-ing Russia as a logical consequence of the historical process of the 20th century —

_if_ it

really had been about

those constructive solutions and not about a neo_imperi-alist striving of the power elites of the main national states, as it has come up again since the end of the Cold War:

A neo_imperialist striving by the power elites of the main na-tional states — in Europe led by the German EU hegemon.

End of the reflections on
the consequences of Yugoslavia's destruction

Especially _if_ the German side, due to the self-destructive re-sults of the two parts of the Great War had taken the only stand that would have been appropriate:

In future we will exclusively engage in political solutions that are geared in a constructive way.

The _first_ consequence of which would have been

> (__and this would have been indispensably have meant to discarding those ideological blinkers, that eventually rendered the Nazism possible__)

to confidently discard political ideas that nowadays, still have a lasting, more or less unconscious effect, such as the one that the population of a state in the best case would have to be ethnically homogeneous.

And _that_

> would have been the only convincing proof for the claim of having understood and really learned the lessons of the Great War.

Therefore,

> we will

_o n l y_

> support constructive solutions.

Consequently,

> we will only support those solutions, that are in accordance with the interests of the people.

For this is what we owe to our national identity grown from harmful experience.

Certainly, this does not only seem *un*_realistic, but it actually would have been unrealistic. But why would it have been unrealistic? Well,

> just because of the 'mere' fact that the political backdrop is taken by the power elites of the national states, respectively

because

> of the factual aspirations of these elites and their satel-
> lites — especially and including the German power elite.

Those "aspirations" became obvious once again when Milan
Panić ...

> The successful racing cyclist, biochemist, philanthropist,
> successful businessman and politician Milan Panić, who
> was born in 1929 in Belgrade and has now lived in the US
> for long, was the Prime Minister of the Federal Republic
> of Yugoslavia from July to December 1992. Originating
> from the Socialist Federative Republic of Yugoslavia
> which consisted of Serbia and Montenegro from April
> 1992 to February 2003 after the other four constituent
> republics (__Slovenia, Croatia, Macedonia and Bosnia-
> Hercegovina__) had declared their independence.
>   It was Milan Panić who played a crucial part in the _In-
> ternational Conference on the Former Yugoslavia_ (ICFY)
> that was held in London on 26/27 October 1992 and
> therefore is also called the _London Conference_ and which
> was about to find a solution for the conflict in Bosnia-
> Hercegovina that had developed so sanguinarily.[19]

... informally expressed the idea that both the USA and Russia
should jointly deploy forces along the river Drina, separating

---

[19]As it might have become clear from the explanations above, this civil
war is to be understood as one of the early consequences of the irresponsible
policy on the Balkans by the German government under chancellor Kohl —
accompanied and demanded by the German media and the then opposition
parties.

Serbia from Bosnia-Hercegovina — this way separating the sanguinarily conflicting parties.[20]

Bosnia-Hercegovina	
	had always been an entity in the South-Slav association of states of Yugoslavia, even though the hyphen might make you assume otherwise. The people of Bosnia-Hercegovina, besides, mostly of Muslim belief, are not an ethnical group but merely offsprings of the Serbs and the Croats who converted to Islam during the time of the rule of the Ottoman Empire.[21]

A part of the border between the East Roman and West Roman Empire ran along the Drina which is the main reason why east of the Drina The Christian Orthodox belief prevails (__which is today represented by the Serbian Orthodox Church__) while west of the Drina the Roman-Catholic belief does.

But this in principle right idea of Panić's could neither be taken up by the parties involved nor could it be formulated in further details with reference to the entire conflict area on the Balkans[22] — as this would not have been in the sense of the different foreign power elites and the different nationalists on the

---

[20] I dearly do appreciate Professor Robert M. Hayden's kindness to let me have this information on 28 March 2017.

[21] See: Annotation 1 and 4 in: Hayden, "Yugoslavia's Collapse: National Suicide with Foreign Assistance", in: *Economic and Political Weekly*, Vol 27, No 27 (__July 4, 1997__) or online at: http://jstor.org/stable/4398583 (__this link was re-checked on 3 August 2018__).

[22] See the peace plan by Hayden, pp 37-9.

Balkans. Mind you, the ordinary population would probably have welcomed this idea no matter which religion they belonged to in their regions — for what else were they so severely divided by?

Main difference between
Serbs, Croats and Slav Muslims

If you look at it — out-of-the-box —, you can get the impression that the main difference between Serbs, Croats and the Slav Muslims is not of linguistic nature, for they all speak different dialects of one language, even though the Serbs use Cyrillic and the Croats, and the Muslim Slavs use Latin.[23]

> Of course, you can always carry dialectic peculiarities caused by geographical separation such as valleys, mountain ridges or large streams to extremes, and this way emphasise the linguistic independence, preferably by eliminating other elements of the dialect, replacing them through elements of the "own" dialect.

That means the main difference between the demographic groups in the former South Slav republic on the Balkans

(__besides Yugoslavia simply means "land of the South Slavs"__)

originates from children being born into a certain religious community which does not necessitate a personal religious af-

---

[23] See annotation 4, in: Hayden, "Yugoslavia's Collapse: National Suicide with Foreign Assistance".

filiation,

> but first of all it is a matter of habit
> and one that is not self-determined.

Should critical differences be drawn from this, it clearly shows the power of ideologies _and_ why these ideologies do exist at all:

> They are based on the problem that their creators and their representatives have with the nature of the human being. After all, the only thing we can know for sure about this

(__unknown [__human__] being__)

is, on the one hand, that its substantial potential is far more than what these creators and representatives of ideological thinking-in-the-box might assume. On the other hand, this unknown being finds its expression in the existing human being in a specific but malleable way.

Man is malleable

QUOTATION

[...] Man as a species is from his intrinsic potential first and foremost: malleable. The way we develop crucially depends on the both social and natural environment. [So] the nature of the single individual human being is shapeable. Mind you, this shapeability decreases with the steady forming progress of the personality, but at the beginning of its development it is only limited through the genetic predisposition, that is to say that a human being could never be, for example, a horse.

The actual forming of the individual personality starts the moment when a human being comes into this world and then this process takes quite a few years. But there is an earlier stage with a completely unconscious process going on. This process takes place in the uterus. [...]

[_Thus,_] the individual human being is neither the one thing nor the other one from predisposition, but it is malleable from its "intrinsic potential" — (__with genetic defects being disregarded for now__). [...]

The societal guidelines and the initial complete plasticity determine the actual shaping, which, once this process is completed, [...] can only experience change through _extreme_ emotional traumata. [...] His perception of the world (__his view of the world__) is according — and so is his perception of the people he encounters [...]. It is in the specific encounter that, on this basis, his own behaviour and the other person's behaviour develops. [...]

Consequently, it is wrong to assume that the system which is to be without alternatives, only exists,

> because the essential

human being be a "moral freak":

> "greedy" _and_ "evil" _and_ "aggressive".

A personality that has been shaped in such a way _can_ show these features, however, it is a fact that the "system" only exists as its apologists, and amongst others its political representatives have put egoism first. But this is a deliberate choice.

[...]

_Man is egoistic!

_No, man is altruistic!

_No, man is both!

Possible. But what prevails?

Whatever prevails is what has been put first societally.

END OF QUOTATION[24]

End of the annotation: Man is malleable

The designation of the constituent republics of the former Yugoslavia

The designation of the constituent republics in the former Yugoslavia originates from the majority of the people living in that republic, even though there have still lived and live large minorities of the other populations.[25]

---

[24] This quotation is a composition of several passages of The *tri*_logical Dissection [...], volume I.

[25] See: Hayden, „Yugoslavia's Collapse: National Suicide with Foreign Assistance", Annotations 2 and 6 (__also see source citation in footnote 21, page 57__).

The common claim that Yugoslavia

was only an "artificial state"

consisting of people that were fundamentally hostile and incompatible to each other

and therefore, alone could not have endured

is to be rejected.

For,

according to Hayden's line of reasoning which is one that looks at things without blinkers,

founding Germany, Italy, India, Belgium or Switzerland was just as artificial and,

objectively seen

there are no more differences, if not fewer, between Serbs and Croats than, for example, between Prussians and Bavarians in Germany when it comes to language, culture and religion.[26]

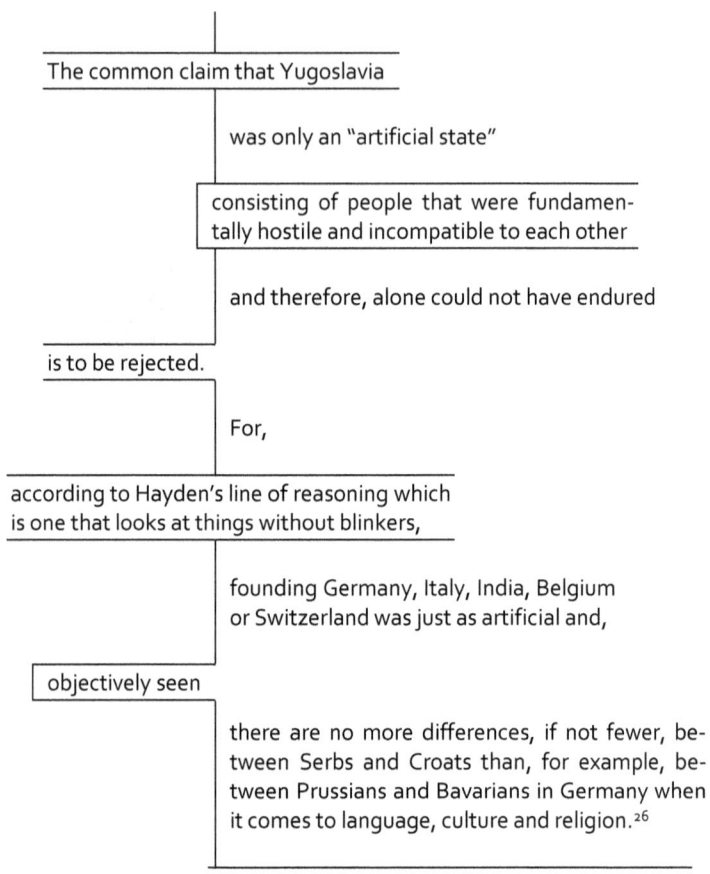

[26] See idem, cit. loc., page 1378.

For the sake of completeness, it be noted that the foundation of the state of Yugoslavia in 1918

(_resulting from the nationalities question still being un-solved just as it was before the first part of the Great War_)

was a result of the first part of the Great War and goes back to the endeavor of the South Slavs, that means of the Serbs, Cro-ats and Slovenians living there,

(_whereas in _this_ context it would not make sense to mention the Muslim part of the population, as they are not a national entity, but basically, they are Serbs and/or Cro-ats merely belonging to the Islamic community of faith_)

to establish their own state as they had obviously realised that they had far more in common than the different dialects of the same common Serbo-Croatian language and the mere fact of being born into different communi-ties of faith might divide them.

That means, eradicating of the both Serbian and on the other side the Croatian elements of the Serbo-Croatian language, that started in the 1990ies, was a socio-patho-logical expression of what is called nationalism.

Thus, in the final analysis,

nationalism serves the purpose not to have the peoples and the mass of the people belonging to different peoples not realise their common interests for, otherwise, the power elite's opportunities and possibilities to manipulate, executed by their satellites in politics, *spin_*doctorial sciences and the media would drastically shrink.

* * *

Another quintessential example the German irresponsible politics is crucially responsible for is the war of aggression against Serbia that was started in 1999 by the leading states of the West. The following quotation is suitable to demonstrate that, as it exemplarily shows the factitiousness and hypocrisy practised to maintain or implement power interests.

The war of aggression against Serbia was the beginning of the era of the "wars to protect the human rights"

QUOTATION:

When in March 2014 due to the obsequies the Serbian people also commemorated the fifteenth anniversary of the aggressive war of NATO against their country from 24 March to

10 June 1999, when many children, women and men fell victim to the airstrikes flown by the so-called defense alliance, to the 1,300 cruise missiles or to the 37,000 cluster bombs dropped, which also destroyed bridges, hospitals, schools, houses, power plants and industrial plants,

"coincidentally"

nothing could be heard or read about that grief in the western mass media.

For

how could this fit in with all the talking about the end of the European "peace culture" by the,

so it was claimed,

Russians breaking international law when they followed the referendum on the Crimea on 16 March 2014

at least it took place,

and probably prevented a bloodbath, when they had the (__Russian__) troops,

that had already been deployed there,

surround the Ukrainian barracks so that this referendum, to which western observers had been invited, could take place unimpeded?

Believing that such an observing was not legitimate because this referendum was "against international law" is questionable firstly because at that time the West had installed a regime in Kiev that was anything but democratically legitimised. And as two recent surveys show the Ukraine of 2016 is a country torn in any respect.[27]

As this applies for all the countries where the neoliberal elites want to see their ideas of a society in line with the market to be installed.[28]

[...] Not even an apology could be heard from, for example, those figures that today know well that as of late a "willkommenskultur"

(__which can hardly be translated as "culture of welcoming", for that sounds a bit strange, doesn't it?__)

is supposed to be in Germany.

---

[27] One of these two surveys originates from the Center for Insights in Survey Research that is not suspected of leaning to the left: http://www.iri.org/sites/default/files/wysiwyg/2015_11_national_over-sample_en_combined_natl_and_donbas_v3.pdf. (__The link was re-checked on 3 August 2018.__)

The other survey comes from the Gallup opinion research institute: http://www.gallup.com/poll/187931/ukrainians-disillusioned-leader-ship.aspx?version=print. (__This link was also re-checked on 3 August 2018.__)

[28] The false argumentation regarding the referendum on the Crimea, especially by German politicians, spin_doctorial scientists, the journalists involved and the representatives of the media, will be dealt with later on.

The term "willkommenskultur"

> seems to stem from people, who divide "culture" into different categories but this kind of „culture" is not credible because the whole thing starts with wrong politics which makes people leave their home countries.

Whoever

> does not want to change the wrong politics or is not aware that it was, has been and is primarily western politics that has made North Africa, Syria, Afghanistan and Ukraine prey of more or less concealed warlords can go hang with their "willkommenskultur", which is, as far as I am concerned, only a big show, targeted at foreign countries.

I am convinced that this term is only used to disguise our own actually practised lack of culture.

> This is what the political acting proves.

Needless to say, this criticism is not aimed at all those who have sincerely done their best to care for human beings who had lost their homes, for it is them who are misused if any politician talks about "willkommenskultur", which, in fact, is mere window-dressing."

Ushering this "new" kind of war, i.e., to wage this first "war to protect the human rights" did not only require a new guise for the NATO but particularly one for the German foreign policies. This was necessary in order to sell such policies _without_

(__at least optically not ignorable__)

protest

(__for this probably would have made the majority of the population express their non_willingness to such a policy__),

without much ado, i.e. to sell a war that violates the German constitution as a "robust humanitarian mission" to the public remaining speechless —

thus, as usual, they pretended that they did not feel comfortable in doing so, but in feigned tears, however, so they said, it was indispensable, because, so these politicians explained, it is Auschwitz, that requires that "we", the Germans do not stand apart, when it comes to the slaughtering of the noble, libertarians of the Kosovo Liberation Army (__KLA__) committed by the stone-hearted Serbs,

at least this is what the majority of the journalists of the so-called quality media

(__according to their own statement__)

and know-it-all politicians told the callous audience —

to be precise this is how they lied to them —

of course without any populism ...

And it was not an easy job for a certain Mr Joseph Fischer

> (__a former street fighter who was at that time the German Foreign Minister and a member of the Green party__)

and a Mr Rudolf Scharping

> (__then the Defence Secretary and member of the SPD__),

just to name two of them.

In order to sell this unconstitutional war of aggression, almost without any friction, as a "robust humanitarian operation" to the audience that remained speechless, the one and very party everybody strongly _believed_ to be linked to the peace movement

> The Green Party,

took the same guise.

And this way, for the time being, the public did not realise that, in order to wage a war of aggression against Serbia, fabricated stories had to serve as genuine truth, such as the one that because of what happened in Auschwitz intervening was mandatory. Once these fabricated stories had made the stunning audience believe that the Serbs had committed atrocities, the Albanian KLA, however, had committed themselves to fight for freedom and therefore

> (__without a mandate by the UN__),

the NATO,

that is

"the military extension of amnesty international", as a certain Mr Ulrich Beck[29] expressed it in such an alarmingly cynical way, hereby disclosing the proper sense of the "peaceful western world",

that is the dreadful way of thinking of its representatives —

before

they get down to their dreadful deed,

which is always on behalf of their "international law" but interpreted in their sense and interest ...

... thus the NATO had to wage a war of aggression against Serbia (__although officially it was never called like that__):

because of Auschwitz,

because of the human rights,

because of the international law.

END OF QUOTATION[30]

---

[29] German politician and member of the Green party.

[30] See the German version of The tri_logical Dissection [...], volume I, pp. 333-8.

Another result, directly linked to politics practised this way and being determined by the satellites of the German power elite in Europe, is the establishment of the Mafia state Kosovo in 2008.

> Which shows once again, what kind of person the ones responsible are when it comes to politics and science and that they are of the same ilk as those people have who pursue such policies _here_ in Europe, i. e. they advocate such policies by forming opinions in the according direction, while they in their own wretchedness claim to be on the side of anything "humane", of anything "democratic", therefore they actually are on the side of what is called a society in a market-compliant sense which, of course, is shaped according to the religion-like neoliberal doctrine, the only thing politically right, and they strictly want others to do their bidding — even though it is all about is pursuing the same policies that only create chaos in the other countries, too, as long as in their opinion it sufficiently serves their own striving for power.

In this context, the following is worth quoting — as it is exemplaric.

QUOTATION

[...] Here I would like to get back to Kosovo, for it exemplarily shows the fatal effects of politics applied by politicians who think in nationalistic terms thus at the service of the corresponding power elites. By breaking Kosovo out of the territorial integrity of Serbia, the succession state of the Federal Republic of Yugoslavia, something evolved that can be called a "mafia state". This way, Gert Weißkirchen, the then-spokesman on foreign affairs of the SPD's fraction of the German Bundestag, expressed his incomprehension of the recognition of Kosovo by the German government on 20 February 2008. Similarly, Johannes Jung, the "expert on the Balkans" for the same party, wondered how Kosovo could possibly ever become a "viable state".[31]

And I would like to point out that the one-sided recognition of the independence of Kosovo in 2008 created a precedent, so, if other peoples rightly referred to thereafter, objectively, this claim cannot be denied, by _those_ politicians responsible who awarded this right to the Kosovo-Albanians hereby, in fact, violating resolution "1244"[32] of the UN-Security Council of 1999 and who did _not_ merely "interpret" the resolution "in a biased way", as Georg Nolte, professor for Public Law, International Law and European Law with the Ludwig-Maximilian-University in Munich stated in a guest contribution on FAZ.NET of 13 February 2008, "that biased interpretations of resolutions of the Security Council" would get the nature of a precedence, which could in other contexts also turn on "the West". [...] He also stated that this would outweigh a "short-term relief of a regional tense situation" as a result of such an interpretation. As an example, he quoted Georgia.[33]

However, for *the declaration of independence* of the Crimea summoning such a "precedence" was probably not necessary, for the secession, the referendum and the accession of the Crimea (_then accepted by Russia_) was exactly no "annexation" — no matter how hard the hypocrites present their outrage.[34]

---

[31]See the related Internet link, re-checked on 4 August 2018: http://www.spiegel.de/spiegel/vorab/a-537278.html.

[32] See here: https://peacemaker.un.org/sites/peace-maker.un.org/files/990610_SCR1244%281999%29.pdf, there the annex 1, "point" (_horizontal line_) 6 on p5 and the annex 2 on pp6-7. (_The indicated link was re-checked on 4 August 2018._)

[33] See on FAZ.Net of 13 February 2008: "Kein Recht auf Abspaltung", (_"no right for separation"_): http://www.faz.net/aktuell/politik/ausland/f-a-z-gastbeitrag-kein-recht-auf-abspaltung-1515789-p2.html?printPagedArticle=true; this link was likewise re-checked on 4 August 2018.

[34] See the article by Reinhard Merkel: "Die Krim und das Völkerrecht: Kühle Ironie der Geschichte", first published on FAZ.NET on 7 April 2014, in the appendix I of the German version of The *tri_logical Dissection* [...], volume I.

In contrast, the separation of Eastern Ukraine,

> should the Russian population in Ukraine feel threatened by those extreme right-wings, who were lifted into the saddle by the "West" and who want to realise their ideas of an "ethnically cleansed" Ukraine,

is comparable to the above-mentioned precedence that was created by the Western world.

Both is possible

> should the situation in Ukraine keep developing precariously[35],thus eventually in Russia itself, too — the historical responsibility for which will be in the hands of those agitators in the "West", especially the top politicians of the Green Party who will have to bear historical responsibility.

And so, the flows of refugees in 2015 that became visible for the general public in 2015 and that was caused by wrong western politics applied for a long time seems like the "sheet lightning" of a globally spreading political process of chaotisation.

But since it is Georg Nolte's view that the interpretation of the UN resolution 1244 was a biased one, he is wrong, for it was a deliberate misinterpretation.

After all, this resolution is exclusively about finding a political solution for the crisis in Kosovo, respectively about an "interim framework agreement", so that on the one hand "a substantial self-rule for Kosovo" becomes feasible, _but_ only within the scope and by maintaining the "principles of sovereignty and the territorial integrity of the Federal Republic of Yugoslavia".[36]

---

[35] See in: German version of The tri_logical Dissection [...], volume I, p334, beginning with: "Zu meinen, daß ein solches Beobachten ...".

[36] See the reference in footnote 32 on p72.

That means recognising the independence of Kosovo is a sheer viola-
tion of this resolution of the UN security council. Getting excited about the
separation of the Crimea is typical western hypocrisy especially because the
separation of the Crimea can absolutely be reasoned historically even without the precedence of Kosovo whilst the separation of Kosovo could only be
reasoned with a higher birth rate of Kosovo-Albanians, which is, however,
not the topic of this UN resolution.

> (__After all, the Crimea already had an auton-
> omous status in the times of the USSR. [...]__)

[...]

END OF QUOTATION[37]

* * *

Well, these examples of policies applied in Europe show exemplarily that these policies cannot be the ones of the common
interests of the European peoples, as all these examples show
that they were not only severe political mistakes but an expression of imperialist policies that go back to the time before

---

[37] Apart from the present *Reading Nine*, see notably the readings 6 and
8 of the English version of The *tri*_logical Dissection of the Lobbycratic Era,
volume III: "'I do not agree!' Politico-economic Readings about the *Neo*_Wil-
helmo-liberalism and its Consequences" from which, as you know, the present special volume *Reading Nine* is a part.

the first part of the Great War.[38] As a consequence, the German policy, starting with the end of the Cold War, is to be called *neo*_Wilhelminist. — And that includes both the specific German nationalist and the old imperial thriving hidden in the German policies _applied_.[39]

End of the excursion:

Quintessential examples of counterproductive consequences
of the German machtpolitik

In other words:

Brussels serves the "indirect game" of machtpolitik.

Whoever wants to make others believe that there has been only one single decision in Brussels that was not desired by the power elites of the national states, does obviously not want to realise the fact that there has always been a fundamental discrepancy between the interests of the mass of people and those of the power elite of the corresponding national state.

Mind you, once upon a time democracy was recognised to be the right form of rule to eliminate such discrepancy.

---

[38] See also in the German version of The *tri*_logical Dissection [...], volume II, Zwischenruf 28.

[39] See notably reading 8 in the English version of The *tri*_logical Dissection [...], volume III.

But the problem is that there has not been democracy any more long-since. In this respect the often heard remark that people be disenchanted with democracy is pure cynicism — for how could they be so if it were about _their_ interests?

However, there is something that is sold as democracy but actually is lobbycracy and, as a consequence, people are disenchanted with lobbycracy.

So, Brussels only serves the famous political "indirect game" that is so common in machtpolitik.

That means I am through with

the neoliberal elite whose members portray themselves as European. Furthermore, I do not want to have anything to do with them, so I have no more truck with them —

the responsibility for anything else is in their hands.

Ever since the beginning of the 1990ies, politics has never aimed at drawing real conclusions from history, which was not only desirable but also feasible.

On the contrary, today we have reached a point where this monstrous EU construct can never ever lead to

(_representative_) democracy — once the current lobbycratic implementation process has been finished.

Whereas if the indispensable factor of democracy were present,

videlicet transparency on all political decision-making levels,

it would not allow for this "elite" to take any step in the direction of this politics.

Or even, this neoliberal elite would not exist at all.

But as this transparency is lacking, creating a lot of ballyhoo suffices to distract the mass of people to implement the particular interest of a few.

Therefore, it is so ridiculous that, for example, any neoliberal person utters, even with a feigned look full of cares for „our" democratic values respectively, the „western values", that the big mistake in the TTIP negotiations was that they were not conducted in a sufficiently transparent manner.

(_?_)

But how should that be possible?

After all, we are talking here about negotiating a major agreement that would objectively be beneficial to relatively few people. Thus, how can you possibly negotiate and draw up any agreement with full transparency if it is not useful for the mass of the people? For "full transparency" would actually mean negotiating and drawing these agreements up, available to everybody, thus

during

the process of negotiation, e.g. on the internet.[40]

---

[40] See the annex of the English version of The tri_logical Dissection [...], volume III, which is about such an agreement named CETA.

That means this entity, that is to be called a real existing lob-bycracy can only make the conditions for the mass of people go to the worse with every further step of its political en-trenchment —

> this is what the neoliberal _dictatorial_ man-ner has significantly proven so far.

This manner, that can for instance be seen in the reactions to the Greek bailout referendum in 2015 and that was shaped by absolutely inacceptable policies from both the side of the "Troika"

> consisting of the ECB, the EU commission and the IMF

> with the European hegemon based in Berlin eventually directly or indirectly having set the direction just as he is doing today,

and from the side of the Tsipras government[41].

Or it can be seen in those airs that were shown in the run-up to and following the Brexit-referendum — just remember words such as:

> "We, the Europeans, now stand together!", uttered by German politicians — after this referendum ...

To my mind, there is something Dadaist-surreal about one of _those_ people trying to tell me: "We, the Europeans!". I mean those people that are changing Europe into a monstrosity that

---

[41] Cf. the German version of The tri_logical Dissection [...], volume II, Zwischenruf 6.

is dangerous for the entire world — and that are using tales supposedly meant to create identity, freshly edited by their pen-pushers but following the same old pattern: "The good guys here — the bad guys there" ...

followed by the fight against them.

So that they then can spread the news that for example a Mr Putin or a Mr Trump be the causal problem.

For, while policies by Putin, in-deed,

are the result of wrongly conducted poli-cies ever since the end of the Cold War,

Trump is the up-dated symptom

of the whole problem of the system.

And it be repeated,

for it is exemplarily for the whole system:

The CETA referendum of the Walloons (__this region in Belgium, home to 3 million people__) was called "anti-democratic" by German politi-cians and those journalists involved of the media concerns, although it was the exact opposite and, in a way, a shining hour of democracy.[42]

[42]See in the mentioned German version of The *tri_logical Dissection* [...], Zwischenruf 11, pp173-4.

Quotation:

If lobbycratic politicians claim that it has nothing to do with democracy when the people in one region realise that a trade agreement is incompatible with democratic rules, and that means

after

these people did what _so far_ had not been done in any other region in Europe and that is having a close look into such an agreement and its impact and consequences

for them,

and then, having done so,

come to the conclusion,

that such an agreement (__of course!__) must be rejected,

then

these lobbycratic politicians call this _anti_democratic and _anti_European.

Consequently,

politicians whose doing, namely turning us into objects of powerful single interests

(__which again has _nothing_ to do with "free trade"__)

are inacceptable for _mature_ thus _enlightened_ people!

END OF QUOTATION[43]

Ratiocinatively it can be said that the representatives of the real-existing lobbycracy know only one direction, which will, at the latest with the establishment of a European army lead to a *neoliberal_military_dictatorial* form of rule, which, to say it explicitly, is to be understood as what I call *neo*_Wilhelmoliberalistic.[44]

In other words, the neoliberal policy delivers those political monstrosities, that it then pretends to be fighting — and the result is: The removal of the remnants of the democratic rights.

---

[43] See at the same place, p. 174.

[44] See reading 8 of the English version of The *tri*_logical Dissection [...], volume III.

ANNEX

Evidence for the allegation that
the EU is an antidemocratic entity

Agreements such as TTIP are not ratified by the members of parliament if there are real representative democratic conditions.[45]

> Mind you, the names of those agreements are absolutely irrelevant for they are not at all about „free trade ", all they are about is the fixation of the protection of investments, that can be claimed according to special law — which per se is incompatible with both democracy
>
> (__there is no special law in democracy__)
>
> and market economy
>
> (__because it eliminates the risk for entrepreneurs__).
>
> So, agreements à la TTIP are to be rejected in a representative democratic society in principle.

However, the fact that the members of parliament do ratify those agreements is _objectively_ seen a proof that we have already been living in a lobbycracy.

---

[45] TTIP stands for "Transatlantic Trade and Investment Partnership".

Should any member of parliament wish to deny this accusation „vehemently",

> believe me,

it would not matter just because of the state of facts.

> For one thing, the negotiations take place in private hearings, for another thing with _direct_ participation of the lobbyists of private economic interests, while the members of parliament not only do not have any say in or impact on the content but they also do not really know what they are eventually going to take a vote on, let alone the consequences it will have on society — consequences not only for the present generation but also for the future generations, often falsely cited in other contexts.

All these aspects, whose content will be looked into in the following.

The mere fact that the text[46] of the CETA (__as an agreement à la TTIP__) was only made available in the mother tongues of the contracting partners once everything was signed, sealed and delivered proves me right.

---

[46] CETA stands for "Comprehensive Economic and Trade Agreement".

Excursive explanations on the CETA agreement

Since the completion of the draft of this agreement à la TTIP in February 2014 there has been an English version of the CETA agreement online.[47]

Brexit-relevant annotation

With reference to the relation between the EU dominated by the German hegemon the question arises in how far the German machtpolitik is eager to deny Great Britain access to the EU market if the Brexit is executed, after all, Great Britain can have access via Canada, which Great Britain is connected to through the Commonwealth. But at another place CETA says that the entry for "other states" be generally amenable — thus, if it was not primarily about striving for dominance, an interim solution would be easy to find until an agreement for the post-Brexit period, constructive for both sides, is found.

Furthermore, a kind of 'revised' version could serve as an interim solution for the time being whilst new regulations for those states wishing to leave the EU are being drafted. This, of course, would bear a certain irony, if a 'revised' version of CETA were the base for maintaining comprehensive trade relations — when the Euro zone collapses.[48]

End of this annotation

---

[47] https://endemannverlag.com/wp-content/uploads/2018/06/tra-doc_CETA_en.version.pdf.

[48] See in the German version of The *Tri_logical* Dissection [...], vol. III, part 4, Reading 17, there the annotations on pp560-63 starting with: "natürlich müßte und könnte es sofort Sonderregelungen geben".

It was not before July 2016 that the text of the agreement was made available in the various mother tongues of the EU member states, i.e. only after the text could not be altered by the parliaments any more.

> Once the agreement is signed,
>
> however, amendments can be made, but exclusively by the neoliberal, mind you, legitimised through free elections, Canadian government or by the explicitly neoliberal EU commission which, on the contrary, is not democratically legitimised, as article 30.2, paragraph 1 of chapter 30, CETA states:
>
> *The Parties may agree, in writing, to amend this Agreement. An amendment shall enter into force after the Parties exchange written notifications certifying that they have completed their respective applicable internal requirements and procedures necessary for the entry into force of the amendment, or on the date agreed by the Parties.*[49]

Whether these "amendments" must be approved by the respective national parliaments and the EU parliament, is not clearly stated in the text, so that it would be conceivable that those amendments that are considered "unfavourable" for the private clients of the lobbyists are made past the public.

QUOTATION:

[...] So why not conclude an agreement, for instance CETA, some articles of which are _for the time being_ and — to calm down the public — judicially

---

[49] See CETA, Chapter Thirty, article 30.2, paragraph 1 on page 226.

ordered to be "abeyant", thus not "applicable", but

> (__somehow on the QT activated__)

these articles do apply nonetheless — following further lobbyist efforts to dispose people? [...]

> END OF QUOTATION[50]

On the other hand, nothing may be altered among other things, with the rules for the investment protection:

> *Notwithstanding paragraph 1, the CETA Joint Committee may decide to amend the protocols and annexes of this Agreement.*[...] *This procedure shall not apply to* [...] *amendments to the annexes of Chapters Eight (__Investment__), Nine (__Cross-Border Trade in Services__), Ten (__Temporary Entry and Stay of Natural Persons for Business Purposes__) and Thirteen (__Financial Services__). [...]*[51]

End of these excursive explanations on the CETA agreement

As indicated above there is not a single explicit word in the agreement text that the EU parliament be granted a special right to check the version of the agreement ratified on 15 February 2017 in an appropriate way _in advance_ — which would absolutely not even have been sufficient according to demo-

---

[50] See in the German version of The *tri*_logical Dissection [...], vol. II, Zwischenruf 18, on p241 the annotation starting with: "Wieso also nicht ein Abkommen abschließen, bspw. CETA [...]".

[51] See CETA, loc. cit., par. 2.

cratic minima standards. But this fact also shows that the decision-makers lack a real democratic awareness, for the content of this agreement has a major impact on the interests of the mass of the people.

> Therefore, CETA is the result of the wheeling and dealing of the EU commission, the according national neoliberal governments of the EU member states and the neoliberal Canadian government as well as the lobbyists involved.

The Canadian government at least have their own experience with the "North American free trade agreement NAFTA" at their command, so they ought to know what such an agreement means.

> After all, in 1994 this agreement between Canada, the USA and Mexico served and still serves as a blueprint for all the following contracts of this kind.[52]

In other words:

> CETA is to be understood as an example for the wheeling and dealing disguised as some international law on behalf of the power elites and it represents a, however well-disguised, but direct attack on the democratic rights of the mass of the people — in this case affecting the people both in Canada, and in the Europe of the EU.

---

[52] See in German on YouTube the following link which was re-checked on 4 August 2018: „NAFTA — Freihandelsabkommen oder Blaupause des neoliberalen Investitionsregimes": https://youtu.be/Gz48jb2_gzc; and, regarding a NAFTA like agreement, read in English: "The Free-Trade Regime: Oligarchy in Action — Free trade agreements like the Trans-Pacific Partnership undermine democracy and sovereignty" by Morris Sanchez:
https://fpif.org/free-trade-regime-oligarchy-action/, this link was checked on 4 August 2018, too.

From 2009 to 2014 CETA was negotiated

> _closed to the public_

but with

> _permanent_

presence of the lobbyists of the powerful economic interests, and on 30 October 2016 it was signed by the EU council, the EU commission and the Canadian government, after the neo-liberal governments of the EU member states had agreed.[53] On 15 February 2017, the EU parliament ratified this agreement, so that on 21 September 2017 those parts that solely affect EU right, came into effect —

> although the individual national parliaments had not ratified it yet.

Admittedly, having the full CETA agreement come into effect is indeed going to depend on how long it will take the national parliaments to vote on it, but when we look at the way these agreements are negotiated and applied, everybody now must be aware of who has all the rights and who can enforce them and, in contrast to that, who is meant for gradually living a life of maids and menials in the globalising lobbycracy, respectively the mass of the people in the so-called democracies of the western world who are supposed to be pleased that they are allowed to serve as maids and menials in one of the concern states of the lobbycratic system — and these "servants"

---

[53] See in the German version of The _Tri_logical Dissection [...], vol. III, part 3, Reading 10, there the annotations on pp256-59 starting with: "... wer in der EU _anti_demokratisch ist ...".

then are convinced that they are far better off than those maids and menials, at whose expenses we live, in the so-called developing countries with all the wealth of their resources serving the system —

as far as they are not collectively fleeing their home countries.

This way the politicians "responsible" determine that future generations will not have the chance any more to freely shape "their" societies — for woe betide them if they did, for shaping their society would violate the profit interests of the investors in "their" countries, woe them for such a shaping violated the profit interests of the investors in "their" countries, even regarding those profits for which no investment has been done at all so far

(__thus, these profits are not real, they are speculatively calculated__),

nevertheless, they would be cause for claims for damages — which government, if there then were still be democratically elected governments, would still dare to go for political ideas and projects that would be *primarily* in the best interest of the general wellbeing, but not in the best interest of profit-making?[54]

---

[54] See also in German: „Das Investitionsschutz-Kapitel im EU-Kanada-Freihandelsabkommen (CETA): Eine kritische Analyse", meaning: „The article for the protection of investments in the EU_Canada free trade agreement (CETA)" by Peter Fuchs, on:
https://power-shift.de/wordpress/wp-content/uploads/2016/04/PowerShift-Analyse-ISDS-in-CETA_Fassung23-April-2016.pdf this link was rechecked on 4 August 2018.

Well, it cannot be denied any more: There can only be a

globally

concerted revolution against this with these agreements fully established lobbycracy — which, of course would be of a totally different nature of that reactionary revolution, whose instructions can be found in the dogma of the neoliberal ideology[55] which is currently globally and concertedly taking place. To say this very clearly: Should this _real_ revolution not happen, the degeneration of the human race is not effectively to be halted.[56]

---

[55] See in the German version of The tri_logical Dissection [...], vol. I, part 1.

[56] In this context please also refer to the German version of: loc. cit, vol. III, part 4, appendix I, pp854-5 starting with: "Das Verbrechen der Digital-Politiker" (__meaning: "The crime of the 'digital politicians'"__). Besides, in The tri_logical Dissection [...], vol. I, part 4: "Der Lösungsweg", chapters 24-5, I have developed a solution that starts from the status quo of the lobbycratic society. This solution is based exclusively on reforms. Mind you, on the contrary, the remark in the text above refers to a firmly revolutionary process that takes place _globally_ concerted. This process will be written out in the final volume of my main book project that is supposed to start in 2019.

But even if the members of parliament had had enough time, their factual understanding of the real meaning of CETA and its consequences would not have been ensured. That means they would probably not really have understood what CETA is:

It is the copestone of the lobbycratic system.

For those texts are only to be understood by those who actually formulated them, stuck in their own trains of thoughts which could exclusively be understood on condition that you also enter into these trains of thoughts.

Yet it would have been a different matter, if Members of the European Parliament (__MEPs__) had had a chance to be present at the drafting of the text, thus, each chapter and article would have been discussed with independent experts — only then, they,

as representatives of the people (__?__),

would have been able to grasp what they actually declared valid on 15 February 2017.

Given the circumstances, that have become in a way creepingly lobbycratic, however, members of parliament cannot apprehend what they have caused by agreeing, this is only possible to those who are really familiar with those contract texts.

> The more so as the special kind of formulations is going to motivate the imagination of any lawyer working for a major enterprise to find the according claim regarding "investment protection" that is always interesting for an enterprise, as those claims are always and _exclusively_ about money.

The CETA Tribunal

Furthermore, it is only up to the "CETA court" called "the Tribunal" to interpret the formulations of the agreement. This so-called court normally consists of 15 members, (__5 from the EU, 5 from Canada and 5 from a third country__), appointed by the CETA Joint Committee[57], comprising representatives of the European Union and representatives of Canada, co-chaired by the Minister for International Trade of Canada and the Member of the European Commission responsible for Trade, or their respective Designee, which in turn means that they are not appointed by the investors' side, as initially envisaged — which is considered to be progressive.

> But on whose side are the neoliberal governments that hardly have the majority of the votes in elections, thus lack the support of the mass of the people[58], respec-

---

[57] See CETA, Chapter Eight, article 8.27, paragraph 2 on page 59.

[58] Also see in the German version of The tri_logical Dissection [..], volume III, part 3, reading 10, pp299-305 starting with: "Statement zu dem Appell Gregor Gysis", as well as pp 488-491 the gloss: "Wie den demokratielähmenden Parteienstaat loswerden?", and in: loc. cit., "Schlußwort, Schlußsatz 2: 'Der Gipfel'", pp778-9 starting with: "So bspw. Herr Macron in Frankreich oder Herr Laschet in Nordrhein-Westfalen" and also the appendices V and VI.

tively the EU administration? Who are they advised by if it is not the lobbyists of powerful single interests?

The members of the "Tribunal" are to be "independent" and not to be affiliated to any government, their decisions are to comply with the "International Bar Association Guidelines on Conflicts of Interest in International Arbitration".[59] Until further notice the Members of the Tribunal shall be paid a monthly retainer fee to be determined by the CETA Joint Committee. This way the tendency to "prolong" an investment lawsuit might arise.

> Though the CETA Joint Committee can envision, *by decision*, to *transform the retainer fee and other fees and expenses into a regular salary, and decide applicable modalities and conditions.*[60]

"Appellate Tribunal" and "Investment Lawsuits"

> The paragraphs one and two of the Article 8.28 state that an Appellate Tribunal is to be established whose task is to uphold, modify or reverse a Tribunal's award.

> It should be emphasised that those "investment lawsuits" are always about an investor suing a state or its bodies _exclusively_ for monetary reasons.

Parallel Jurisdiction

> Furthermore, these lawsuits follow, indeed, a special jurisdiction whose "investment laws"

---

[59] See CETA, article 8.30, paragraph 1 on page 63.
[60] See loc. cit., article 8.27, paragraph 15 on page 61.

are yet to be formulated by this "court" (__thus the "Tribunal"__) to be constituted upon CETA coming into force.

This way, a parallel jurisdiction is established, a jurisdiction for which the "principle of equality" does not apply any more, as the agreement has the effect that the "investors" are _more_ equal than the mere states because it is exclusively the "investors" who can sue states and their bodies.

As mentioned above, later on there is to be an "appellate tribunal" where states can appeal a verdict indeed — just as the defeated "investor" can.

"No precedences"

It is also quite interesting that no case at the CETA Tribunal may be taken as precedence for other cases.

Does that sound like making a "'binding' special jurisdiction" possible?

Or does it sound more like comparable cases might end up in "thumbs up" or "thumbs down", as this is a

"flexible special jurisdiction",

just as the neoliberal doctrine postulates —

when it comes to "flexibility"?

You might as well call it "judging at your discretion".

When rendering its decision, the "Tribunal" is to apply the CETA as interpreted in accordance with the *Vienna Convention on the Law of Treaties* (VCLT) that has been in force since 1969 and is itself an international treaty (__See CETA, article 8.31__).

The "Investment court/Tribunal" is a provisional arrangement

> Well, I am sure that this will be music to a neoliberal ideologist's ears, however, for any human being who thinks in societal contexts all this seems highly questionable. And yet all this is only the forceful entering into the lobbycratic era.

It is obvious that this "CETA court" is merely a kind of provisional arrangement at the beginning of the now starting lobbycratic era because in the future it will be replaced by "the establishment of a multilateral investment tribunal and appellate mechanism":

> *The Parties shall pursue with other trading partners the establishment of a multilateral investment tribunal and appellate mechanism for the resolution of investment disputes.* [61]

By that point of time at the latest the end of the bourgeois democratic era will be manifested.

The interpretation of a dispute through this tribunal, which is, it be repeated, exclusively about money, will then be binding and, in case of a lawsuit against a state as a whole or parts of

---

[61] See loc. cit., article 8.29 on page 63.

it, e.g. against a single small town, serves as its base — and the base for this lawsuit will not be formulated in its full extent until the approval of CETA itself has been executed by the national parliaments and the EU parliament.[62]

And it is also true that on the one side there is the, at least elected, yet neoliberal government (__Canada__) that signed this agreement, however, on the other side there is the European Commission that was not elected by the European peoples that preempted the decision and actually only

wanted

to submit the result for "voting" to the EU parliament.

It was only after the European Court of Justice (__ECJ__) intervened that the national parliaments were involved in the ratification: The EU Commission,

(__that had been opposing a participation of the national parliaments in the approval of CETA for a long time__),

wanted to know from the judges whether the trade agreement with Singapore, signed in 2013, could be taken as a precedence that those agreements indeed fall only in the remit of the EU-Commission and the

---

[62] See in German: „Trojanisches Pferd der Konzerne? — Sieben Anmerkungen zu CETA" by Erik Jochum on Makroskop.eu. The following link to this essay was re-checked on 4 August 2018: https://makroskop.eu/2016/11/trojanisches-pferd-der-konzerne-sieben-anmerkungen-zu-ceta/.

> EU-Parliament without involving the national parliaments of the EU member states.

> The „EU Singapore Free Trade Agreement" (__EUSFTA__), is an agreement of the TTIP type, so that a verdict by the ECJ would make it a precedence for agreements like that.

The assessment on this matter by the ECJ of 16 May 2017 comes to the result that the national parliaments are to be included in the approval process in the aspects that have a direct effect on the national interests, thus the national parliaments also have the right of veto, which could even entail the failure of such an agreement.[63]

> Well, apparently none of the politicians responsible that were democratically elected came by themselves up with the idea to call for exactly that as *sine qua non* — thus as something that is an essential precondition for something to be done.

So even if, of course, this decision by the ECJ is to be appreciated it does not really change the fundamentally anti-democratic nature of the EU.

For

> whoever signs such trade agreements,

> that have a deep impact into human

---

[63] See the press release by the ECJ on: https://www.mehr-demokratie.de/fileadmin/pdf/2017-05-16_EuGH-Urteil.pdf; this link was re-checked on 4 August 2018.

societies and this way subjects them to more and more ongoing trimming in line with the market,

and submits the procedure of the whole agreement directly to the lobbyist impact that determines it,

and closed to the public,

and who only then after that procedure submits the agreement for voting to members of parliaments who were also not elected democratically therefore under insufficient conditions

(__after all, this is a work of agreement that the members of parliament have no chance to comprehend because of its way of formulating, as explained above__),

can only command a consciousness _freed_ from democracy.

Thus, here is a good question for you:

How was it possible for the single member of parliament to get to learn and understand the content of this agreement that serves as a special legislative basis?

What sources of information did they have on hand?

If they did not want to be "advised" by (__at least__) one of the lobbyists surrounding them

> *and especially in this case, as it was about the fixation of the lobbycratic rule, it was probably business as usual namely that the members of parliament unsolicited were slipped some "explanatory" material by the lobbyists,*

who, when it comes to this agreement, were and have been as active as they will be, as this is all a lobbyist gets paid for. Did,

> for example,

the members of the German parliament also referred to as *Deutscher Bundestag* have the chance to read the "interpretation" elaborated by the division in charge of this agreement — as far this can actually be stated because the only thing that was really "interpreted" is the crucial points in chapter 8 and chapter 21?[64]

Annotation:

> The architrave of the facade of the "Reichstag" which is the building that hosts the German parliament, bears an inscription: "Dem deutschen Volke". This inscription is often misinterpreted, although the meaning is quite clear, provided, you look at it unconfused, for "Dem deutschen Volke" does not mean: "The German People" but something like: "Given

---

[64] This report of 29 August 2016 by the „Fachbereich Europa" of the German Bundestag: „Kapitel 8 und 21 des CETA-Abkommens — Regulierungszusammenarbeit und Investitionsschutz", is available on: https://www.bundestag.de/blob/476042/4e706fdac1d463d97654998b157b85e5/pe-6-127-16-pdf-data.pdf; this link was re-checked on 4 August 2018.

to the German people", expressing that the "Reichstag" was passively given to the German people, consequently this inscription implies that the German people did not take it in a collective self-liberating action. The problem here is two-fold: The first one is the ignorance regarding the true meaning of the inscription: "Dem deutschen Volke" and the second one is the underlying reason "to give" a building to the German people as their parliament, a reason that is an undemocratic one. Consistently the Wilhelminist Kaiser Wilhelm called the Reichstag "talking shop". In fact, this description was and is true, for the rulings were and are indeed not made in this parliament but in closed back-rooms. The CETA is just an example.

End of this annotation

If you read these „interpretations of CETA", your skepticism will be reinforced, especially as the authors apparently tried hard to inform the so-called representatives of the people "in an unbiased way" .

But what does „unbiased" actually mean when it is about an agreement on the same level as international law, yet hostile to democracy, that in practice is applicable to two parties — the entity of a polity on the one side and the entity of particular economic interests that is ideologically reasoned on the other side, whereby, in addition, the latter side holds the whip hand, too?

The „right to regulate"

Formulations such as the

„right to regulate"

and

„investment protection"

should, at least, worry, particularly when they are used in the same context.

At least for the authors of this „interpretation of the agreement" (_remember the annotation in footnote 64 on page 100_) the term „right to regulate" seems to be crucial, for it is mentioned 35 times in this "paper".

After all, I have the impression that the term „right to regulate" is a formulation that was admittedly taken into the text because of the public pressure. However, in Chapter Eight, Section D of the CETA: "Investment protection", Article 8.9: "Investment and regulatory measures", the term „for greater certainty", despite being used twice does not bring that clarification that would have avoided a dispute — between an investor and a state — in the first place. On the contrary formulations of that kind virtually ask for clarification ...

at the "special court" called *investment tribunal*.

Or alternatively, such formulations (_e.g., "right to regulate" / „for greater certainty"_) have the effect that the politicians responsible are not even going to change the societal parameters to

the benefit of the mass of people in the territory of their own
country

(__despite having promised to do so in an election c_r_ampaign__)

— for fear of a lawsuit of an „investor" and the consequent potential compensation as it could be seen umpteen times.[65]

It is said that it has already happened in Canada
that a company _successfully_ sued a com-mu-
nity because the community's democratic decision _would have_ caused a loss in profit if the
company _had set up_ a production plant there.

To make it clear it be repeated:

We are talking about a loss in profit that _would have_ happened

(__thus, calculated on pure speculation__)

if this company _had set up_ a production plant
in this community, which they have never done.

---

[65] Source in German: „Materialsammlung: Fallbeispiele zu Konzern-
klagen gegen Staaten"; the following link was re-checked on 4 August 2018:
https://www.ttip-stoppen.at/wp-content/uploads/2014/03/fallbei-
spiele_Schiedsgerichte_final-neu.pdf; and also, in: *Blätter für deutsche und
internationale Politik*, „CETA: Blaupause der Deregulierung"; the following
link was re-checked on 4 August 2018, too: https://www.blaetter.de/ar-
chiv/jahrgaenge/2015/februar/ceta-blaupause-der-deregulierung?print.

Against the background that has been common practice for long, formulations such as those in chapter 23 more or less seem to be a parody on the actual conditions if, e.g., you just look at the work conditions in Germany. For, in the 2000s a huge low page sector was established including legally underpaid contact workers and a legal minimum wage that is merely a cosmetic one — let alone the social means of gagging called HARTZ IV.

Excursion

The term HARTZ IV is named after the former German manager Peter Hartz who was chief human resources officer and member of the *VW Group* management board and dismissed in July 2005, due to the corruption affair within the *VW Group* and who gave his name for this part of the so-called *Agenda 2010*, the so-called *labour market reforms*

(__*En passant, the official name of this means of gagging is: "Volume II of the (German) Social Code" that got his first draft on 24 December 2003, to come into effect on 1 January in the following year, so that it could get its full validity on 1 January 2005.*__)

This concept of Peter Hartz's

(__*incidentally, he has later on distanced himself from elements of this* "approach to 'reform'"__)

conceals what it actually means:

A social gagging firstly directed against those people who are in work under fragile circumstances due to the ongoing competition anyone against anyone, so each of them needs to have the slogan on the lips: *How can I sell myself best?* O t h e r w i s e he will get more and more pressurised and consequently drifts off his usual social level down to the level of HARTZ IV in other words: he is caught in the „HARTZ IV cage" because the options to get out of this "cage" again are excessively restrained.

Now, these so-called reforms are serving as a model for all the members of the European Monetary Union (__EMU__).

Explanatory annotation:

One of the main flaws of the EMU is shown by the lack of a replacement for the no longer existing currency adjustment between countries made through revaluation and devaluation of their currencies due to the introduction of a single currency in the EMU.

(__The revaluation and devaluation *of currencies* determine the international competitiveness of an economy [__*shown in the exchange rate of its currency*__] and the relative level [__*of its exchange rate of its currency*__] seen over the years, allows to read the purchasing power of that economy — o n c o n d i t i o n that there are no speculative attacks against such a currency that lead to a distortion of the purchasing power so that the whole _real_ economy of such a country would come under a non-self-induced pressure, in other words, initiated by such a foreign exchange speculation.__)

Thus, it should be obvious that a monetary union needs a replacement for this "valve of revaluation and devaluation

(__of currencies__)" common between all
market societies _o u t s i d e_ such a union. This "valve of replacement" is just the
"Golden Rule of Wages". (__A definition of
this rule can be found in Reading One.__)

End of this excursion

Dear Reader, it is up to you to find a proper name for this entity called the EU.

"Civil Society Forum"

In articles 22.1-5 of chapter 22 headed: "Trade and sustainable
Development", the reader is also told about a so-called "civil
society", whereby article 22.5 is even called: "Civil Society Forum". Well, the organisations participating in this so-called
Civil Society Forum —

> like independent representative employers, unions, labour
> and business organisations, environmental groups, as well
> as other relevant civil society organisations as appropriate[66]

---

[66] See in the English version of the CETA, paragraph 2 of the article
22.5 on page 183.

> (__"*other relevant civil society organisations as appropriate*" —
> what could be meant by that? Perhaps a kind of so-called NGO like
> the now only exemplarily mentioned NGO "Pulse of Europe"?__),

have internalised the neoliberal credo for long, and function
accordingly in the neoliberal sense in line with the market,

> as they prove at any public occasion.

So, you cannot tell me that the consequent results of discussions that took place in the "Civil Society Forum" can be anything but what those omnipresent lobbyists have been recommending on behalf of "independent, representative employers" as well as the corresponding "employers' associations".

QUOTATION

Affirming the value of greater policy coherence in decent work, encompassing core labour standards, and high levels of labour protection, coupled with their effective enforcement, the Parties recognise the beneficial role that those areas can have on economic efficiency, innovation and productivity, including export performance. In this context, they also recognise the importance of social dialogue on labour matters among workers and employers, and their respective organisations, and governments, and commit to the promotion of such dialogue.

END OF QUOTATION[67]

---

[67] See paragraph 2 of article 23.1 on page 184 in the English version of the CETA. Well, you can gain an insight of what it means in practice when "social partnership dialogues" are talked about in the German version of The *tri*_logical Dissection [...],, Volume III, part 2: "Von neoliberaler Ideologie, marktkonformen Arbeitnehmervertretern und einigen exemplarischen Konsequenzen" (__meaning: "About the neoliberal ideology, workers' representatives who are in line with the market and some exemplary consequences"__), pp182-237.

Maybe you also want to look at chapter 24: "Trade and Environment", as another example?

Well, right here it should be sufficient to remind you of the reality of reoccurring food scandals such as animal feed infested with dioxin, pesticides in beer, antibiotics in pork, beef and chicken, rotten meat that had already been distributed to the consumers or eggs infested with the insecticide *Fipronil*, all of which are in the end systemic.

And all these scandals have one thing in common:

The appropriate action has never been taken — for it is the market that regulates everything.

From an ideological point of view this phrase:

"it is the market that regulates everything" is definitely consistently mouthed, nevertheless you could also come to the conclusion that a way of production

(__that is not hedged in by the Social State of Law[68], thus merely__)

geared at making profits, will sooner or later turn into mafia style.

---

[68] The "Social State of Law" should not be confused with the bourgeois "social state", regarding the differences see the German version of The *Tri*_logical Dissection [...], Volume I, part 4, chapters 23-5.

Therefore, the mentioned chapters 22, 23 and 24 quoted from the CETA

have to

be a parody, for if you especially wanted to take the content of chapter 23: "Trade and Labour" at face value, the German entrepreneurs would vehemently have to speak out against this agreement, but they do not.

1,598 pages of patient paper

Well, they do not need to, after all labour policies like the ones mentioned above have been negotiated by the so-called „social partners", haven't they?[69]

That means, this agreement in its English version consists of 1,598 pages of patient paper, too.

Wouldn't one sheet (__perhaps printed on both sites__) for the contract have been enough, if hardly any state will dare to file a lawsuit at a

regular

court against an „investor", because this investor violated contrac-tual working standards and — conditions through his production methods or because this investor damaged the environment?

At least this would make other potential „investors" look at one an-other in silent agreement, if the democratically elected government of a country only dared to discuss the possibility of suing such an „investor" — at the national labour court of that country — perhaps even dared to discuss the matter with the affected citizen priorly.

---

[69] See in the English version of: loc. cit., volume III, Reading Eight: "The Europe of the neo-Wilhelminist Hegemon and its Future".

A certain „mentality" that is inherent in those „investors" can be deduced from the way the EU deals with people who dare to hold a referendum against projects, that the mass of the people can only perceive as threatening — and that are formulated in contracts à la TTIP, can't it?

> Mind you, how will they be dealt with once the EU army has been established — as well as the appertaining surveillance system, which will be far more efficient than the one of the Stasi of the former GDR?

Not to mention a _democratic_ government would

really

dare to sue against profit-oriented projects.

You know, what would happen in that case?

> Well, they would simply create a party with a reliable party leader — just like the French did in April 2016 with the artificial product "En Marche!" created particularly by the *Cap Gemini Group* and the corresponding leading figure, namely Emanuel Macron, who they now can both admire and sustain[70] — and although in the Presidential election in 2017 this figure got actual 16% of all the votes, the neoliberal elite called that a "landslide victory".

---

[70]See on « Challenges.fr », of 12 April 2016 the article : „Ceux qui s'agitent derrière la start-up Macron", whose link was re-checked on 4 August 2018: https://www.challenges.fr/politique/ceux-qui-se-cachent-derriere-la-start-up-macron_29092.

Example of stories told in contrast to the facts

Mr Macron did win the French election, but based on what? Well, in the second round of voting on May 7, 2017, 64 % of "his" voters elected Mr Macron to prevent Ms Le Pen.

Nonvoters: about 25 %

(__In 2012 the participation was 80 %, according to *Le Monde* this has been the lowest one in a final ballot ever since 1969.__)

Vote blanc
(__voters go to the polls but do not vote for a candidate__): about. 8.5 %

Void: about 3 %

That leaves about 88.5 % of actual votes for either candidate. Of these 88.5 percent about 66 % voted for Macron and about 34 % for Marine Le Pen, which makes it actually 44 % of all the eligible voters (__47,568,693__) who voted for Macron, and about 22 % for M Le Pen.[71]

If you now take into consideration that 64 % of the Macron voters did not vote for him because he was their first choice but only to prevent a Le Pen government, that leaves 36 % of the Macron voters for Macron.[72]

---

[71] Source: The French Ministry of the Interior:
https://www.interieur.gouv.fr/Elections/Les-resultats/Presidentielles/elecresult__presidentielle-2017/(path)/presidentielle-2017/FE.html; this link was re-checked on 4 August 2018.
[72] Source: Screenshot of 7. May 2017:
https://pbs.twimg.com/media/C_PSFn2WsAEQ95x.jpg; this link was re-checked on 4 August 2018, too.

In other words, this president actually has a "back-ing" of barely 16% among the French population.

The same applies to his political movement, which is now called "La République en Marche": In the first round of voting on 11 June 2017, this artificial political product reached about 15 % of the votes cast, with an electoral turnout of less than 50 %.

This second strange result was also celebrated by all propaganda media as an "overwhelming victory!".

Wherefore do we need satire, when in the lobbycratic era satirical reality is as comprehensive as it is "good"?

Termination [of contract]

To conclude I would like to draw your attention to paragraph 2 of article 30.9 of the CETA on page 229 which says that

(_theoretically_)

there is the possibility to terminate a contract with an enter-prise, however this enterprise is entitled to claim for payment of compensation up to twenty years

after

the termination ...

* * *

Looking at the elucidation in this additional reading it is to be concluded that contracts à la TTIP (__e.g. CETA__) are

> _not_

negotiated by people who think in terms of democracy and not in lobbycratic dimensions.

| Thus, the EU is a lobbycratic project. |

And, as a consequence, whoever argues in favour of this EU,

> _objectively_

places himself in opposition to the mass of the European population. The same applies for the representatives, elected by this population:

> If these representatives sign an agreement, that assumedly will make enterprises sue whole parts of the population in their states,

> (__just because of the ambiguous and contradictory formulations that need interpretation by entering an action — and that always need to start from scratch again as there are no precedences in this "special court"__),

> they have exposed themselves as being on the side of those whose aim is, amongst others, the abolishment of

| (__representative__) |

democratic conditions. This will be especially the case as these lawsuits will be at a „tribunal", that is yet to decide based on a special law that needs interpretation — and the defendants will always be states. And all this against the backdrop of the

> _well-known_

negative impacts of agreements already signed, such as NAFTA, the *North American Free Trade Agreement* between the USA, Mexico and Canada, an agreement that can be seen as forerunner.[73]

And as if that were not enough, even if one contracting party could free itself of the strait jacket, that means would be able to terminate the contract, it be repeated: the "protection rules for investors" remain valid for 20 more years to come.[74]

Besides —

a real person loses their legal status as entrepreneur if that person wants to have their investments contractually suable protected, even those investments that have not even been made, this way maximising their profits — although the condition for the right for a certain amount of profit has to be that this goes hand in hand with taking the full entrepreneurial risk.

In conclusion, the following be quoted:

QUOTATION

That means my assumption that [...], by so-called free trade agreements à la TTIP, they want to create a global means in order to have a direct impact on the national economies within the

---

[73] See indications in the footnote 52 on page 88.
[74] See CETA, Chapter 30, article 30.9, paragraph. 2 on page 229.

ambit of such agreements would have to be proven wrong before I can be convinced that these agreements actually served
the purpose of "free trade" because the contents of such agreements are drafted by lobbyists of the globally operating concerns (__thus, eventually by the power elites__), and not by independent experts (__who have the comprehensive general over-
view that is necessary in those contexts__). So, these agreements
can only serve the protection of the interests of globally operating concerns, not the protection of humane societies as
such.[__p302__]

In practice, this is the phenomenon that is called imperialism
— and that is today just as present as it was before the first part
of the great imperialist war. And in that context, too, secretly
negotiated free trade agreements à la TTIP between the USA
and the EU or between blocks and countries elsewhere in the
world are to be understood as follows:

> Maintaining the national states only in the sense of a
> „replacement for feeling home" for the *peoples that
> are nationally organised,* with simultaneous full de-
> velopment of the concerns, not only _without_ any
> chance for impact for the national states but also
> _against_ the interests of the peoples organised in
> the national states — if you dared to cut the oppor-
> tunities for profit for the concerns.[pp350-1]

The fact, that they say it was hard-nosed post-negotiating from
the side of neoliberal-oriented politicians that changed the term
(__"free trade arbitration tribunals"__) to investment courts
merely shows that it is about blurring an

_u n a m b i g u o u s l y_

antidemocratic project, for substantially _nothing_ has changed in
the neoliberal thrust, even though this may be considered necessary and is in the consequence of the neoliberal ideology — if you
look at this development from the neoliberal perspective.

Thus, the advocates of those free trade projects just do not tell the truth when they claim that it be a mistake if those negotiations take place behind closed doors for: how could such negotiations be run in a transparent way? And nobody who thinks democracy is not just a mere phrase can accept that a small group of people secure themselves special rights and have complete populations dance to their pipes, which is the case exactly because of those so-called free trade agreements

(__and absolutely corresponds to the neoliberal thinking__)! [pp481-2]

END OF QUOTATION[75]

* * *

It may well baffle you but there are no credible studies proving the claim the "free trade" per se be good and globally foster economic growth. This belief is especially deeply anchored in the neoliberal ideologues' minds so that they do not even need those studies.[76] That is not only worrying but it is politically irresponsible that, based on ideological assumptions, a world order is going to be anchored institutionally via treaties à la TTIP and, by so doing, they are initiating a collective degeneration process.

* * *

[75] The three parts of this quotation can be found in the German version of The tri_logical Dissection[...], Volume I, part 3, there on pp302, 350-1 and on 481-2.

[76] See in: loc. cit., volume III, part 3, Reading Twenty, pp675-81, starting with: "Neoliberale Ideologen behaupten ...".